PROCEEDINGS
of the
STEKLOV INSTITUTE
of
MATHEMATICS

edited by

I. G. Petrovskiĭ and S. M. Nikol'skiĭ

number **88** (1967)

APPROXIMATION OF FUNCTIONS IN THE MEAN

edited by

S. B. Stečkin

AMERICAN MATHEMATICAL SOCIETY

Providence, Rhode Island

1969

Академия Наук
Союза Советских Социалистических Республик

ТРУДЫ МАТЕМАТИЧЕСКОГО ИНСТИТУТА
имени В. А. Стеклова
LXXXVIII

Ответственный редактор академик И. Г. ПЕТРОВСКИЙ
Заместитель ответственного редактора профессор С. М. НИКОЛЬСКИЙ

ПРИБЛИЖЕНИЕ ФУНКЦИЙ В СРЕДНЕМ

Сборник работ
Под редакцией С. Б. Стечкина

Издательство „Наука“
Москва 1967

Translated from the Russian by
R. P. Boas and H. R. Jaschke

Printed in the United States of America

PROCEEDINGS OF THE STEKLOV INSTITUTE OF MATHEMATICS
IN THE ACADEMY OF SCIENCES OF THE USSR

(Труды математического института им. В. А. Стеклова, т. LXXXVIII, 1967)

TABLE OF CONTENTS

JACKSON'S THEOREM IN L_p

V. I. BERDYŠEV

Let L_p be the space of functions of period 2π, integrable together with their pth powers ($1 \leq p < \infty$), with the norm

$$\|f(x)\|_p = \left(\int_{-\pi}^{\pi} |f(x)|^p \, dx \right)^{\frac{1}{p}};$$

let

$$E_n(f, L_p) = \inf_{T_{n-1}} \|f(x) - T_{n-1}(x)\|_p$$

be the best approximation to $f(x) \in L_p$ by trigonometric polynomials $T_{n-1}(x)$ of order at most $n-1$; let

$$\omega(t, f, L_p) = \sup_{|h| \leqslant t} \|f(x+h) - f(x)\|_p$$

be the integral modulus of continuity of $f(x)$; let C be the space of continuous functions of period 2π with norm $\|f(x)\|_C = \max_{|x| \leq \pi} |f(x)|$; let $E_n(f, C)$ be the best approximation to $f(x) \in C$ in the C metric by trigonometric polynomials of degree at most $n-1$; and let $\omega(t, f, C)$ be the modulus of continuity of $f(x)$.

Jackson's theorem [1] states that

$$E_n(f, C) \leqslant K\omega\left(\frac{\pi}{n}, f, C\right) \qquad (n = 1, 2, \cdots), \tag{1}$$

where K is an absolute constant. N. P. Korneĭčuk [2] proved that the smallest possible constant K in (1) is 1; more precisely, he proved that

$$\frac{2n-1}{2n} = \sup_{f \in C} \frac{E_n(f, C)}{\omega\left(\frac{\pi}{n}, f, C\right)} < 1 \qquad (n = 1, 2, \ldots).$$

The analogous problem for approximation in the L_p metric ($1 \leq p < \infty$)[1] has not been solved. It is well known (see for example [3]) that

$$E_n(f, L_p) \leqslant K\omega\left(\frac{\pi}{n}, f, L_p\right) \qquad (n = 1, 2, \ldots), \tag{2}$$

where K is an absolute constant, but the smallest possible value $K = K(p)$ in this inequality is unknown.

In §1 we give a lower bound for the best constant in (2). A central role in the discussion of this problem is played by a lemma on measurable sets

[1] On the case $p = 2$ cf. the paper by N. I. Černyh [6] in this volume.

that was established by A. M. Il'in and is published here for the first time, with his permission. In §2 we consider a similar problem for approximation by linear operators of convolution type.

§1.

Let E be a set on $[0,1]$ and let $\widetilde{E}$ be its continuation to the real axis with period 1. Let $\widetilde{E}_t$ be the set of points of the form $x+t$, $x \in \widetilde{E}$, where t is real, and let E_t be the set $\widetilde{E}_t \cap [0,1]$.

Lemma 1 (A. M. Il'in). *Let* $0 < m < 1$. *For each* $\epsilon > 0$ *there is a set* $E \subset [0,1]$, *with* $\operatorname{mes} E = m$, *such that*

$$\operatorname{mes} E \cap E_t \geqslant m^2 - \varepsilon. \tag{3}$$

for each t $(0 \leqq t \leqq 1)$.

Proof. It is enough to prove the lemma for $m \leqq \frac{1}{2}$, since in the contrary case we can consider the complementary sets CE and CE_t, and apply the equation

$$\operatorname{mes}\,(CE \cap CE_t) = 1 - 2 \operatorname{mes} E + \operatorname{mes} E \cap E_t. \tag{4}$$

Let k be an integer and $q < m$. Later we shall take k large and q small.

We proceed to construct a set $E = E(q) = E(q,k)$ of measure m. We construct inductively intervals of two kinds, indexed from 0 to N. The integer N will be specified later. We call $[0,1]$ an interval of the first kind of index 0, and $[1-q,1]$ an interval of the first kind of index 1. We then dissect $[0,1-q]$ into k equal parts and call each of the resulting intervals an interval of the second kind of index 1. We treat the intervals of the second kind in the same way as we treated $[0,1]$. That is, let $[a,b]$ be an interval of the second kind of index r, $1 \leqq r < N$. Then we call $[b-(b-a)q, b]$ an interval of the first kind of index $r+1$. We divide $[a, b-(b-a)q]$ into k equal parts and call each of them an interval of the second kind of index $r+1$. The sum of the lengths of all intervals of the first kind up to index N (inclusive) is $q\sum_{r=1}^{N}(1-q)^{r-1}$. We determine N by

$$q\sum_{r=1}^{N}(1-q)^{r-1} \leqslant m < q\sum_{r=1}^{N+1}(1-q)^{r-1},$$

i.e.

$$1-(1-q)^N \leqslant m < 1-(1-q)^{N+1}. \tag{5}$$

There is always such an N since $0 < q < m < 1$ and $1-(1-q)^{r+1} \to 1$ as $r > \infty$.

We now construct E. We include in E all the intervals of the first kind of index up to and including N. In addition, in each interval of the second kind of index N we take an arbitrary set of measure $(1/k^N)\{m - q\sum_{r=1}^{N}(1-q)^{r-1}\}$ and include it in E. This is possible since the length of an interval of the second kind of index N is $((1-q)/k)^N$. Since the number of intervals of the second kind of index N is N^k, we find that $\operatorname{mes} E = m$.

We now find a lower bound for $\operatorname{mes} E \cap E_t$. Let α^r be any interval of the first kind of index r. Let A^r be the union of the intervals of the first kind of index less than r. We dissect $E \cap E_t$ into N subsets, $E \cap E_t = \bigcup_{r=1}^{n} B_r$, where $B_1 = (\alpha_t^1 \cap E) \cup (\alpha^1 \cap E_t)$ and B_r $(r = 2, 3, \cdots, N)$ is the union of the sets $\bigcup_{\alpha_t^r \cap A^r = 0} (\alpha_t^r \cap E)$ and $\bigcup_{\alpha^r \cap A_t^r = 0} (\alpha^r \cap E_t)$; here $\bigcup_{\alpha^r \cap A_t^r = 0}$ denotes the union of the intervals of the first kind of index r that do not intersect A_t^r. Evidently

$$\bigcup_{r=1}^{N} B_r \subset E \cap E_t, \qquad B_i \cap B_j = 0 \quad (i \neq j;\ i,\ j = 1,\ 2,\ \ldots,\ N),$$

and hence

$$\operatorname{mes} E \cap E_t \geqslant \sum_{r=1}^{N} \operatorname{mes} B_r. \tag{6}$$

To estimate $\operatorname{mes} B_r$ $(r = 1, 2, \cdots, N)$, we count the number of intervals α^r that satisfy $\alpha^r \cap A_t^r = 0$. It is easily seen that $\operatorname{mes}\{[0,1] - (A^r \cup A_t^r)\} \geqq (1-2q)^{r-1}$, and the number of intervals of the first kind of index less than r is $n_r = \sum_{l=1}^{r-1} k^{l-1}$. Each interval α^r belongs to just one interval of the second kind of index $r-1$. The length of an interval of the second kind of index $r-1$ is $((1-q)/k)^{r-1}$. Hence the number of intervals of the first kind of index r that satisfy $\alpha^r \cap A_t^r = 0$ is not less than $\bar{n}_r = [(1-2q)^{r-1}(k/(1-q))^{r-1}] - 2n_r$. Similarly the number of intervals α_t^r for which $\alpha_t^r \cap A^r = 0$ is not less than $\bar{n}_r$. Thus the set B_r $(r = 2, 3, \cdots, N)$ consists of at least $\bar{n}_r$ subsets of the form $\alpha^r \cap E_t$ and $\bar{n}_r$ subsets of the form $\alpha_t^r \cap E$.

Let $r = 1$. If $\alpha^1 \cap \alpha_t^1 = 0$, the interval α_t^1 intersects at least $[qk/(1-q)]$ intervals of the second kind of index 1. Each interval of the second kind of index 1 contains a subset of E of measure $(m-q)/k$, and hence

$$\operatorname{mes} \alpha_t^1 \cap E \geqslant q \frac{m-q}{1-q} - \frac{1}{k};$$

and similarly

$$\operatorname{mes} \alpha^1 \cap E_t \geqslant q \frac{m-q}{1-q} - \frac{1}{k}.$$

Hence

$$\operatorname{mes} B_1 \geqslant 2q \frac{m-q}{1-q} - \frac{2}{k}.$$

We now estimate $\operatorname{mes} B_1$ when $\alpha^1 \cap \alpha_t^1 \neq 0$. Evidently

$$(\alpha_t^1 \cap E) \cup (\alpha^1 \cap E_t) = \{(\alpha_t^1 - \alpha^1) \cap E\} \cup \{(\alpha^1 - \alpha_t^1) \cap E_t\} \cup \{\alpha_t^1 \cap \alpha^1\}.$$

The set $\alpha_t^1 - \alpha^1$ intersects at least $[(\operatorname{mes}(\alpha_t^1 - \alpha^1)/(1-q))k]$ intervals of the second kind of index 1. Hence

$$\operatorname{mes}\{(\alpha_t^1 - \alpha^1) \cap E\} \geqslant \operatorname{mes}(\alpha_t^1 - \alpha^1) \frac{m-q}{1-q} - \frac{1}{k}.$$

In the same way

$$\operatorname{mes}\{(\alpha^1 - \alpha_t^1) \cap E_t\} \geqslant \operatorname{mes}(\alpha^1 - \alpha_t^1) \frac{m-q}{1-q} - \frac{1}{k};$$

and since $m \leqq \frac{1}{2}$ we have

$$\operatorname{mes}(\alpha^1 \cap \alpha_t^1) \geqslant 2 \operatorname{mes}(\alpha^1 \cap \alpha_t^1) \frac{m-q}{1-q}.$$

Combining the last three inequalities, we obtain

$$\operatorname{mes}\{(\alpha_t^1 \cap E) \cup (\alpha^1 \cap E_t)\} \geqslant (\operatorname{mes}\{(\alpha_t^1 - \alpha^1) \cap E\} + \operatorname{mes}\{(\alpha^1 - \alpha_t^1) \cap E_t + \operatorname{mes}(\alpha^1 \cap \alpha_t^1)) \frac{m-q}{1-q} - \frac{2}{k} = 2q \frac{m-q}{1-q} - \frac{2}{k}.$$

Hence we have proved that

$$\operatorname{mes} B_1 \geqslant 2q \frac{m-q}{1-q} - \frac{2}{k}.$$

Let r, $1 < r \leqq N$, be arbitrary. Suppose that α_t^r does not intersect any intervals of the first kind of index r. Then α_t^r has a nonempty intersection with at least $[qk/(1-q)]$ intervals of the second kind of index r. Each interval of the second kind of index r contains a subset of E of measure $m - q\sum_{l=1}^{r}(1-q)^{l-1}/k^r$, and therefore

$$\operatorname{mes}(\alpha_t^r \cap E) \geqslant \frac{q}{k^{r-1}} \frac{m - q\sum\limits_{l=1}^{r}(1-q)^{l-1}}{1-q} - \frac{1}{k^r}. \tag{7}$$

It is easily verified that if the interval α^r does not intersect any interval of the first kind of index r, shifted by t, then

$$\operatorname{mes}(\alpha^r \cap E_t) \geqslant \frac{q}{k^{r-1}} \frac{m - q\sum\limits_{l=1}^{r}(1-q)^{l-1}}{1-q} - \frac{1}{k^r}. \tag{8}$$

As in the case $r = 1$, it is easy to see that if α_t^r intersects an interval $\bar{\alpha}^r$ of the first kind of index r (it can intersect at most one such interval), then

$$\operatorname{mes}\{(\alpha_t^r \cap E) \cup (\bar{\alpha}^r \cap E_t)\} \geqslant 2 \frac{q}{k^{r-1}} \frac{m - q\sum\limits_{l=1}^{r}(1-q)^{l-1}}{1-q} - \frac{2}{k^r}. \tag{9}$$

Applying (7), (8), or (9), depending on whether or not the interval under consideration intersects the specified intervals of the same index, we obtain

$$\operatorname{mes} B_r \geqslant 2\bar{n}_r \left(\frac{q}{k^{r-1}} \frac{m - q\sum\limits_{l=1}^{r}(1-q)^{l-1}}{1-q} - \frac{1}{k^r} \right) \quad (r = 1, 2, \ldots, N),$$

where

$$\bar{n}_r = \left[(1-2q)^{r-1} \left(\frac{k}{1-q} \right)^{r-1} \right] - 2 \frac{k^{r-1} - 1}{k-1}.$$

Using these estimates in (6), we obtain, after elementary transformations,

$$\operatorname{mes}(E \cap E_t) \geqslant 2q \sum_{r=1}^{N} \frac{m - q \sum_{l=1} (1-q)^{l-1}}{(1-q)^r} (1-2q)^{r-1} - \frac{AN}{k} \tag{10}$$
$$= 2(m-1)\left\{1 - \left(\frac{1-2q}{1-q}\right)^N\right\} + 1 - (1-2q)^N - \frac{AN}{k}$$

where A is an absolute constant. By (5),

$$\lim_{q\to 0} (1-q)^N = 1 - m.$$

Hence

$$\lim_{q\to 0} (1-2q)^N = \lim_{q\to 0} (1-2q)^{\frac{1}{2q} \frac{2q}{\ln(1-q)} \ln(1-q)^N} = e^{2\ln(1-m)} = (1-m)^2$$

and

$$\lim_{q\to 0} \left(\frac{1-2q}{1-q}\right)^N = 1 - m.$$

Hence, putting $k = N^2$, we obtain

$$\varliminf_{q\to 0} \operatorname{mes} \{E(q) \cap E_t(q)\} \geqslant 2(m-1)\{1-(1-m)\} + 1 - (1-m)^2 = m^2.$$

Thus for each $\epsilon > 0$ there is a $q < m$ such that

$$\operatorname{mes} \{E(q) \cap E_t(q)\} \geqslant m^2 - \varepsilon.$$

This completes the proof of the lemma.

Remark. By calculating

$$\int_0^1 \operatorname{mes} E \cap E_t \, dt = \int_0^1 \int_0^1 \varphi_E(x) \varphi_E(x+t) \, dx \, dt,$$

where $\varphi_E(x)$ is the characteristic function of E, we can easily show that the conclusion of the lemma is false when $\epsilon = 0$.

We have the following proposition.

Lemma 2. *Let l be an integer, $l \geqq 2$, $\epsilon > 0$, and let the numbers m_i $(i = 1, 2, \cdots, l)$ satisfy $m_i > 0$, $\sum_{i=1}^{l} m_i < 1$. Then there are sets E_i on $[0,1]$, $\operatorname{mes} E_i = m_i$ $(i = 1, 2, \cdots, l)$, such that $E_i \cap E_j = 0$, for all i and j $(i \neq j;\ j = 1, 2, \cdots, l)$, and for each t $(0 \leqq t \leqq 1)$*

$$\operatorname{mes} \{(E_i)_t \cap E_j\} \leqslant m_i m_j + \varepsilon. \tag{11}$$

Proof. We carry out the proof for $l = 2$, the only case that will be used here. For $l > 2$ the proof is similar.

Let $0 < q_1 < m_1$, $q_2 = m_2 q_1 / m_1(1 - q_1)$, and let k be an integer. Later we shall take q_1 small and k large. We proceed to construct $E_1 = E_1(q_1, k)$ and $E_2 = E_2(q_1, k)$. We construct sets of types D_i^j and d_i^j $(j = 1, 2;\ i = 1, \cdots, M)$ by induction. The integer M will be specified later. We call $[0, 1]$ an interval of type D_1^1, and $[1-q_1,1]$ an interval of type d_1^1. Divide $[0,1-q_1]$ into k equal parts and call each part an interval of type D_1^2. Let $[a,b]$ be an interval of type D_1^2. The interval $[b-q_2(b-a),b]$ is taken to be of type d_1^2, and the interval $[a,b-q_2(b-a)]$ is divided into k equal parts, each of which is taken to be of type D_2^1. We proceed similarly with each interval of type D_1^2. We divide intervals of type D_2^1 in the same way as we divided $[0,1]$.

Let $[\alpha_1\beta]$ be any interval of type D_r^1. We call the interval $[\beta-q_1(\beta-\alpha),\beta]$ an interval of type d_r^1, and we divide $[\alpha,\beta-q_1(\beta-\alpha)]$ into k equal parts, each of which is called of type D_r^2. Each interval of type D_r^2 is divided in the following way. Let $[\gamma,\delta]$ be an interval of type D_r^2. We call $[\delta-q_2(\delta-\gamma),\delta]$ an interval of type d_r^2, and we divide $[\gamma,\delta-q_2(\delta-\gamma)]$ into k equal parts, each of which is called of type D_{r+1}^1.

Let l_r^j $(j=1,2;\ r=1,2,\cdots,M+1)$ be the sum of the lengths of the intervals of type D_r^j, let n_r^j be the number of these intervals, and let $\overline{D}_r^j$ and $\overline{d}_r^j$ be the lengths of the individual intervals of types D_r^j and d_r^j. Evidently

$$l_r^1=(1-q_1)^{r-1}(1-q_2)^{r-1},\qquad l_r^2=(1-q_1)\ (1-q_2)^{r-1},$$

$$n_r^1=k^{2(r-1)},\qquad n_r^2=k^{2r-1}$$

and

$$\overline{D}_r^1=\frac{(1-q_1)^{r-1}(1-q_2)^{r-1}}{k^{2(r-1)}},\qquad \overline{D}_r^2=\frac{(1-q_1)^r\,(1-q_2)^{r-1}}{k^{2r-1}},\quad \overline{d}_r^j=\overline{D}_r^j \tag{12}$$

$$(j=1,2;\ r=1,2,\ldots).$$

We determine the integer M by

$$\begin{aligned} q_1\sum_{r=1}^{M} l_r^1 &= q_1\sum_{r=1}^{M}(1-q_1)^{r-1}(1-q_2)^{r-1} \\ &\leqslant m_1<q_1\sum_{r=1}^{M+1}(1-q_1)^{r-1}(1-q_2)^{r-1}=q_1\sum_{r=1}^{M+1} l_r^1. \end{aligned}\tag{13}$$

Then

$$\begin{aligned} q_2\sum_{r=1}^{M} l_r^2 &= q_2(1-q_1)\sum_{r=1}^{M}(1-q_1)^{r-1}(1-q_2)^{r-1} \\ &\leqslant m_2<q_2(1-q_1)\sum_{r=1}^{M+1}(1-q_1)^{r-1}(1-q_2)^{r-1}=q_2\sum_{r=1}^{M+1} l_r^2. \end{aligned}\tag{13$'$}$$

We can rewrite (13) and (13′) in the form

$$1-(1-q_1)^M(1-q_2)^M\leqslant m_1+m_2<1-(1-q_1)^{M+1}(1-q_2)^{M+1}. \tag{13$''$}$$

There is such an M since $m_1+m_2<1$ and $q_1<m_1$.

We now construct the sets E_1 and E_2. In E_1 we put all the intervals of type d_r^1 $(r=1,2,\cdots,M)$. On each interval of type d_{M+1}^1 we insert a set of measure $(m_1-q_1\sum_{r=1}^M l_r^1)/k^{2M}$ and include it in E_1. In E_2 we put all the intervals of type d_r^2 $(r=1,2,\cdots,M)$. On each interval of type d_{M+1}^2 we insert a set of measure $(m_2-q_2\sum_{r=1}^M l_r^2)/k^{2M+1}$ and include it in E_2. The length of an interval of type d_{M+1}^1 is $q_1(1-q_1)^M(1-q_2)^M/k^{2M}$, and the length of an interval of type d^2_{M+1} is $q_2(1-q_1)^{M+1}(1-q_2)^M/k^{2M+1}$; then by (13) and (13') we see that

$$m_1 - q_1 \sum_{r=1}^{M} l^1_r \leqslant q_1(1-q_1)^M(1-q_2)^M,$$

$$m_2 - q_2 \sum_{r=1}^{M} l^2_r \leqslant q_2(1-q_1)^{M+1}(1-q_2)^M,$$

and so the construction is possible. Clearly $\operatorname{mes} E_1 = m_1$ and $\operatorname{mes} E_2 = m_2$.

We now find an upper bound for $\operatorname{mes}\{(E_1)_t \cap E_2\}$. Let A^j_r ($j = 1,2$; $r = 1,2,\cdots,M+1$) be the union of the intervals of type d^j_r and put

$$(B^1_r)_t = (A^1_r)_t - \left(\bigcup_{i=1}^{r-1} A^1_i\right) \cup \left(\bigcup_{i=1}^{r-1} A^2_i\right) \quad (r = 2, 3, \ldots M+1),$$

$$B^2_r = A^2_r - \left(\bigcup_{i=1}^{r} A^1_i\right)_t \cup \left(\bigcup_{i=1}^{r-1} A^2_i\right)_t \quad (r = 1, 2, \ldots M).$$

Then

$$\operatorname{mes}\{(E_1)_t \cap E_2\} = \operatorname{mes}\{(A^1_1)_t \cap E_2 + \sum_{i=2}^{M+1} \operatorname{mes}\{(B^1_i)_t \cap E_2\} + \sum_{i=1}^{M} \operatorname{mes}\{B^2_i \cap (E_1)_t\}. \tag{14}$$

Evidently $\operatorname{mes}\{(E_1)_t \cap E_2\}$ is largest when $A^1_i \cap (A^1_i)_t = 0$, $A^2_i \cap (A^2_i)_t = 0$ for each i ($i = 1, \cdots, M+1$); hence in finding an upper bound for $\operatorname{mes}\{(E_1)_t \cap E_2\}$ we may suppose that this condition is satisfied.

We now find upper bounds for $\operatorname{mes}\{(B^1_i)_t \cap E_2\}$ and $\operatorname{mes}\{B^2_i \cap (E_1)_t\}$. To do this we count the intervals of type d^2_i contained in B^2_i, and compute the measure of the intersection of each such interval with $(E_1)_t$. The measure of $[0,1] - \{A^1_1 \cup (A^1_1)_t\}$ is $(1-2q_1)$. This set intersects at most $\bar{n}^2_1 = [(1-2q_1)/\bar{D}^2_1] + 2n^1_1$ intervals of type D^2_1, each of which contains one interval of type d^2_1. Hence B^2_1 contains at most $\bar{n}^2_1$ intervals of type d^2_1.

The measure of $[0,1] - \{(\bigcup_{j=1}^{i} A^1_j) \cup (\bigcup_{j=1}^{i} A^1_j)_t \cup (\bigcup_{j=1}^{i-1} A^2_j) \cup (\bigcup_{j=1}^{i-1} A^2_j)_t\}$ is $(1-2q_1)^i(1-2q_2)^{i-1}$. This set intersects at most

$$\bar{n}^2_i = \left[\frac{(1-2q_1)^i(1-2q_2)^{i-1}}{\bar{D}^2_i}\right] + 2(n^1_1 + n^2_1 + \ldots + n^2_{i-1} + n^1_i)$$

intervals of type D^2_i, each of which contains an interval of type d^2_i. Hence B^2_i contains at most n^2_i intervals of type d^2_i.

Similarly the measure of $[0,1] - \{A^1_1 \cup (A^1_1)_t\} \cup \}A^2_1 \cup (A^2_1)_t\}$ is $(1-2q_1) \cdot (1-2q_2)$. This set intersects at most $\bar{n}^1_2 = [(1-2q_1)(1-2q_2)/\bar{D}^1_2] + 2(n^1_1 + n^2_1)$ intervals of type D^1_2, shifted by t. Hence B^1_2 contains at most $\bar{n}^1_2$ intervals obtained by shifting intervals of type d^1_2 by t.

The measure of $[0,1] - \{(\bigcup_{j=1}^{i-1} A^1_j) \cup (\bigcup_{j=1}^{i-1} A^1_j)_t \cup (\bigcup_{j=1}^{i-1} A^2_j) \cup (\bigcup_{j=1}^{i-1} A^2_j)_t\}$ is $(1-2q_1)^{i-1}(1-2q_2)^{i-1}$. This set intersects at most

$$\bar{n}_i^1 = \left[\frac{(1-2q_1)^{i-1}(1-2q_2)^{i-1}}{\bar{D}_i^1} \right] + 2(n_1^1 + n_1^2 + \ldots + n_{i-1}^1 + n_{i-1}^2)$$

intervals of type D_i^1, shifted by t. Each of these intervals contains a single interval obtained by shifting an interval of type d_i^1 by t. Consequently the set $(B_i^1)_t$ contains at most $\bar{n}_i^1$ intervals of type d_i^1 shifted by t.

An interval of type d_i^1, shifted by t, intersects at most $[\bar{d}_i^1/\bar{D}_i^2] + 2$ intervals of type D_i^2. Each interval of type D_i^2 contains a subset of E_2 of measure

$$\frac{a_i^2}{n_i^2} = \frac{m_2 - q_2 \sum_{j=1}^{i-1}{}' (1-q_1)^j (1-q_2)^{j-1}}{n_i^2},$$

and hence

$$\operatorname{mes}\{(B_i^1)_t \cap E_2\} \leqslant \frac{m_2 - q_2(1-q_1)\sum_{j=1}^{i-1}(1-q_1)^{j-1}(1-q_2)^{j-1}}{n_i^2} \left(\left[\frac{\bar{d}_i^1}{\bar{D}_i^2} \right] + 2 \right) \bar{n}_i^1 \tag{15}$$

$$(i = 2, 3, \ldots, M+1).$$

An interval of type d_i^2 intersects at most $[\bar{d}_i^2/\bar{D}_{i+1}^1] + 2$ intervals obtained by shifting intervals of type D_{i+1}^1 by t, and each of these intervals contains a set $(E_1)_t$ of measure

$$\frac{a_i^1}{n_{i+1}^1} = \frac{m_1 - q\sum_{j=1}^{i}(1-q_1)^{j-1}(1-q_2)^{j-1}}{n_{i+1}^1}$$

consequently

$$\operatorname{mes}\{B_i^2 \cap (E_1)_t\} \leqslant \frac{m_1 - q_1\sum_{j=1}^{i}(1-q_1)^{j-1}(1-q_2)^{j-1}}{n_{i+1}^1} \left(\left[\frac{\bar{d}_i^2}{\bar{D}_{i+1}^1} \right] + 2 \right) \bar{n}_i^2. \tag{16}$$

We insert the values of n_i^j, $\bar{d}_i^j$, $\bar{D}_i^j$ $(j = 1, 2)$ into (15) and (16) and sum, obtaining

$$\sum_{i=2}^{M+1} \operatorname{mes}\{(B_i^1)_t \cap E_2\}$$

$$\leqslant \sum_{i=2}^{M+1} \frac{a_i^2}{k^{2i-1}} \left(\frac{q_1 k}{1-q_1} + 2 \right) \left(\frac{(1-2q_1)^{i-1}(1-2q_2)^{i-1}}{(1-q_1)^{i-1}(1-q_2)^{i-1}} k^{2(i-1)} + 2\frac{k^{2(i-1)}-1}{k-1} \right)$$

$$= \sum_{i=2}^{M+1} \frac{q_1 a_i^2 (1-2q_1)^{i-1}(1-2q_2)^{i-1}}{(1-q_1)^i (1-q_2)^{i-1}} + \sum_{i=2}^{M+1} \frac{2q_1 a_i^2 (k^{2(i-1)}-1)k}{k^{2i-1}(k-1)(1-q_1)}$$

$$+ \sum_{i=2}^{M+1} \frac{2a_i^2 (1-2q_1)^{i-1}(1-2q_2)^{i-1}}{k(1-q_1)^{i-1}(1-q_2)^{i-1}} + \sum_{i=2}^{M+1} \frac{4a_i^2(k^{2i-2}-1)}{k^{2i-1}(k-1)}$$

$$= \sum_{i=2}^{M+1} \frac{q_1 a_i^2 (1-2q_1)^{i-1}(1-2q_2)^{i-1}}{(1-q_1)^i(1-q_2)^{i-1}} + \Sigma_1' + \Sigma_2' + \Sigma_3',$$

$$\sum_{i=1}^{M} \operatorname{mes}\{B_i^2 \cap (E_1)_t\}$$

$$\leqslant \sum_{i=1}^{M} \frac{a_i^1}{k^{2i}} \left(\frac{q_2 k}{1-q_2} + 2 \right) \left(\frac{(1-2q_1)^i (1-2q_2)^{i-1}}{(1-q_1)^i (1-q_2)^{i-1}} k^{2i-1} + 2 \frac{k^{2i-1}-1}{k-1} \right)$$

$$= \sum_{i=1}^{M} \frac{q_2 a_i^1 (1-2q_1)^i (1-2q_2)^{i-1}}{(1-q_1)^i (1-q_2)^i} + \sum_{i=1}^{M} \frac{2q_2 a_i^1 (k^{2i-1}-1) k}{k^{2i}(k-1)(1-q_2)}$$

$$+ \sum_{i=1}^{M} \frac{2a_i^1 (1-2q_1)^i (1-2q_2)^{i-1}}{k(1-q_1)^i (1-q_2)^{i-1}} + \sum_{i=1}^{M} \frac{4a_i^1 (k^{2i-1}-1)}{k^{2i}(k-1)}$$

$$= \sum_{i=1}^{M} \frac{q_2 a_i^1 (1-2q_1)^i (1-2q_2)^{i-1}}{(1-q_1)^i (1-q_2)^{i-1}} + \Sigma_1'' + \Sigma_2'' + \Sigma_3''.$$

Since $a_i^j < 1$ $(j = 1, 2;\ i = 1, 2, \cdots, M+1)$, there is an absolute constant A_1 such that $\Sigma_1' + \Sigma_2' + \Sigma_3' + \Sigma_1'' + \Sigma_2'' + \Sigma_3'' \leqq A_1 M/k$. Moreover,

$$\operatorname{mes}\{(A_1^1)_t \cap E_2\} \leqslant \frac{m_2}{k}\left(\left[\frac{q_1 k}{1-q_1} \right] + 2 \right) \leqslant \frac{m_2 q_1}{1-q_1} + \frac{2m_2}{k},$$

since $(A_1^1)_t$ intersects at most $[q_1 k/(1-q_1)] + 2$ intervals of type D_1^2, each of which contains a subset of E_2 of measure m_2/k.

Using (14), we therefore obtain

$$\operatorname{mes}\{(E_1)_t \cap E_2\} \leqslant \frac{m_2 q_1}{1-q_1} + \sum_{i=2}^{M+1} \frac{q_1 a_i^2 (1-2q_1)^{i-1} (1-2q_2)^{i-1}}{(1-q_1)^i (1-q_2)^{i-1}}$$

$$+ \sum_{i=1}^{M} \frac{q_2 a_i^1 (1-2q_1)^i (1-2q_2)^{i-1}}{(1-q_1)^i (1-q_2)^i} + \frac{AM}{k} = \frac{m_2 q_1}{1-q_1}$$

$$+ \sum_{i=2}^{M+1} \frac{q_1}{(1-q_1)} \left\{ m_2 - q_2 \sum_{j=1}^{i-1} (1-q_1)^j (1-q_2)^{j-1} \right\} \times \left\{ \frac{(1-2q_1)(1-2q_2)}{(1-q_1)(1-q_2)} \right\}^{i-1}$$

$$+ \sum_{i=1}^{M} \frac{q_2}{(1-2q_2)} \left\{ m_1 - q_1 \sum_{j=1}^{i} (1-q_1)^{j-1} (1-q_2)^{j-1} \right\} \times \left\{ \frac{(1-2q_1)(1-2q_2)}{(1-q_1)(1-q_2)} \right\}^i \tag{17}$$

$$+ \frac{AM}{k} = \frac{m_2 q_1}{1-q_1} + \left(\frac{m_2 q_1}{1-q_1} + \frac{m_1 q_2}{1-2q_2} - \frac{q_1 q_2}{1-(1-q_1)(1-q_2)} \right.$$

$$\left. - \frac{q_1 q_2}{[1-(1-q_1)(1-q_2)](1-2q_2)} \right) \times \sum_{i=1}^{M} \left\{ \frac{(1-2q_1)(1-2q_2)}{(1-q_1)(1-q_2)} \right\}$$

$$+ \frac{q_1 q_2}{1-(1-q_1)(1-q_2)} \left(1 + \frac{1}{1-2q_2} \right) \sum_{i=1}^{M} (1-2q_1)^i (1-2q_2)^i + \frac{AM}{k},$$

where A is an absolute constant.

By (13″),

$$\lim_{q_1\to 0}(1-q_1)^M(1-q_2)^M = 1-m_1-m_2.$$

Hence

$$\lim_{q_1\to 0}(1-2q_1)^M(1-2q_2)^M$$

$$=\lim_{q_1\to 0}(1-2q_1-2q_2+4q_1q_2)^{\left(\frac{1}{2q_1+2q_2-4q_1q_2}\right)\frac{2q_1+2q_2-4q_1q_2}{\ln(1-q_1-q_2+q_1q_2)}\ln(1-q_1-q_2+q_1q_2)M}$$

$$=\lim_{q_1\to 0}(1-q_1)^{2M}(1-q_2)^{2M}=(1-m_1-m_2)^2,$$

since $q_2 = m_2q_1/m_1(1-q_1)$.

Putting $k = N^2$, we obtain

$$\overline{\lim_{q_1\to 0}}\operatorname{mes}\{(E_1)_t\cap E_2\}\leqslant\lim_{q_1\to 0}\left\{\left(m_2q_1+m_1q_2-\frac{2q_1q_2}{q_1+q_2}\right)\frac{m_1+m_2}{q_1+q_2}+\frac{2q_1q_2}{q_1+q_2}\,\frac{2(m_1+m_2)-(m_1+m_2)^2}{2(q_1+q_2)}+\frac{A}{N}\right\}=m_1m_2.$$

The second part of the lemma follows from the equation

$$\operatorname{mes}\{(E_1)_t\cap E_2\}=\operatorname{mes}\{(E_1)_t\cap E_2\}_{1-t}=\operatorname{mes}\{E_1\cap(E_2)_{1-t}\}.$$

This completes the proof of the lemma.

Lemma 3. *Let $0<\bar m<\frac12$ and $0<q<1$. There are subsets $E_1(q)$ and $E_2(q)$ of $[0,1]$, with $\operatorname{mes}E_1(q)=\bar m$, $\operatorname{mes}E_2(q)=\bar m(1-q)$, such that $E_1(q)\cap E_2(q)=0$, and for each t $(0\leqq t\leqq 1)$*

$$\overline{\lim_{q\to 0}}\operatorname{mes}E_1(q)\cap\{E_2(q)\}_t\leqslant\bar m^2,$$

$$\overline{\lim_{q\to 0}}\operatorname{mes}\{E_1(q)\}_t\cap E_2(q)\leqslant\bar m^2, \tag{18}$$

$$\underline{\lim_{q\to 0}}\operatorname{mes}\{E_1(q)\cup E_2(q)\}_t\cap\{E_1(q)\cup E_2(q)\}\geqslant 4\bar m^2.$$

Proof. Let k be an integer. Put $q=q_1$, $\bar m=m_1$, $\bar m(1-q)=m_2$, and consider the sets $E_1(q)=E_1(q_1,k)$ and $E_2(q)=E_2(q_1,k)$, $\operatorname{mes}E_1(q)=m_1$, $\operatorname{mes}E_2(q)=m_2$, constructed in the proof of Lemma 2 (here $q_2=m_2q_1/m_1(1-q_1)=q_1$). It follows from the proof of Lemma 2 that these sets satisfy (17). Taking the limit as $q_1\to 0$ in (17), we obtain

$$\overline{\lim_{q\to 0}}\operatorname{mes}\{E_1(q)\}_t\cap E_2(q)\leqslant\bar m^2.$$

It follows from

$$\operatorname{mes}\{E_1(q)\}_t\cap E_2(q)=\operatorname{mes}E_1(q)\cap E_2(q)\}_{1-t}$$

that $\overline{\lim}_{q\to 0}\operatorname{mes}E_1(q)_t\cap\{E_2(q)\}_t\leqq\bar m^2$.

Since $q=q_1=q_2$, the set $E=E_1(q)\cup E_2(q)$, with $\operatorname{mes}E=\bar m+\bar m(1-q)=m$, coincides with the set $E(q,k)$ constructed in the proof of Lemma 1. Hence E satisfies (10), from which we can obtain the required condition

$$\varliminf_{q\to 0} \operatorname{mes}\{E_1(q)\cup E_2(q)\}_t\cap\{E_1(q)\cup E_2(q)\}\geqslant 4\bar{m}^2$$

by letting $q\to 0$. This completes the proof.

THEOREM 1. *We have the inequality*

$$\sup_{\in L_p}\frac{E_n(f, L_p)}{\omega\left(\frac{\pi}{n}, f, L_p\right)}\geqslant \max\left\{2^{\frac{1-p}{p}}, 2^{-\frac{1}{p}}\right\}\quad (1\leqslant p<\infty). \tag{19}$$

PROOF. We consider two cases.

1) $1\leqq p\leqq 2$. Consider an $\epsilon>0$ and a set $E\subset[0,1]$ with $\operatorname{mes} E=\frac{1}{2}$, satisfying (3). Let $\varphi_E(x)$ be the characteristic function of E and put

$$f_E(x)=f_E(x, E, n)=2\varphi_E\left(\frac{1}{2\pi}nx\right)-1,\quad f_E(x)=f_E(x+2\pi). \tag{20}$$

Evidently $\operatorname{sign} f_E(x)=f_E(x)$, $|f_E(x)|=1$; and

$$\int_{-\pi}^{\pi}|f_E(x)|^{p-1}\operatorname{sign} f_E(x)\cos kx\,dx=0,\quad \int_{-\pi}^{\pi}|f_E(x)|^{p-1}\operatorname{sign} f_E(x)\sin kx\,dx=0$$

for $k=0,1,\cdots,n-1$. Since we also have $\operatorname{mes} E\{f_E(x)=0\}=0$, the polynomial of best approximation to $f_E(x)$ (cf. [4]) is zero, and

$$E_n(f_E, L_p)=\|f_E(x)\|_p=(2\pi)^{\frac{1}{p}}\quad (1\leqslant p\leqslant 2).$$

Moreover, by Lemma 1 we have

$$\begin{aligned}&\|f_E(x+2\pi t)-f_E(x)\|_p\\ &=2\{2\pi[\operatorname{mes} E_t\cap([0,1]-E)+\operatorname{mes} E\cap([0,1]-E_t)]\}^{\frac{1}{p}}\\ &=2\left\{4\pi\left(\frac{1}{2}-\operatorname{mes} E\cap E_t\right)\right\}^{\frac{1}{p}}\leqslant 2(\pi+4\pi\varepsilon)^{\frac{1}{p}}\end{aligned}$$

for arbitrary t $(0\leqq t\leqq 1)$, and hence

$$\omega(t, f_E, L_p)\leqslant 2(\pi+4\pi\varepsilon)^{\frac{1}{p}}\quad (0\leqslant t\leqslant 2\pi); \tag{21}$$

consequently

$$\frac{E_n(f_E, L_p)}{\omega(t, f_E, L_p)}\geqslant 2^{\frac{1-p}{p}}\left(\frac{1}{1+4\varepsilon}\right)^{\frac{1}{p}}\quad (0<t\leqslant 2\pi).$$

2) $2\leqq p<\infty$. Let $\epsilon>0$ and $0<\bar{m}<\frac{1}{2}$. By Lemma 3 there are a number q and sets $E_1(q)=E_1$, $E_2(q)=E_2$, $E_1\cap E_2=0$, $\operatorname{mes} E_1=\bar{m}$, $\operatorname{mes} E_2=\bar{m}(1-q)$, such that

$$\begin{aligned}\operatorname{mes} E_1\cap(E_2)_t&\leqslant \bar{m}^2+\varepsilon,\\ \operatorname{mes}(E_1)_t\cap E_2&\leqslant \bar{m}^2+\varepsilon,\\ \operatorname{mes}(E_1\cup E_2)_t\cap(E_1\cup E_2)&\geqslant 4\bar{m}^2-\varepsilon.\end{aligned} \tag{22}$$

Put

$$\tilde{f}(x)=\begin{cases}1, & x\in E_1,\\ -1, & x\in E_2,\\ 0, & x\in[0,1]-(E_1\cup E_2)\end{cases}$$

and

$$f(x)=f(x,E_1,E_2,n)=\tilde{f}\left(\frac{1}{2\pi}nx\right),\quad f(x)=f(x+2\pi). \tag{23}$$

Clearly

$$|f(x)|^{p-1}\operatorname{sign}f(x)=f(x),\qquad \int_{-\pi}^{\pi}f(x)\cos kx\,dx=0,$$

$$\int_{-\pi}^{\pi}f(x)\sin kx\,dx=0\ (k=1,2,\ldots,n-1)\ \text{and}\ \int_{-\pi}^{\pi}f(x)\,dx=mq.$$

Hence by a well-known argument we obtain $|\,\|f(x)\|_p-E_n(f,L_p)|\leq C\overline{m}a$, where C is an absolute constant. In fact, if $T^*_{n-1}(x)$ is the polynomial of best approximation to $f(x)$ then

$$\int_{-\pi}^{\pi}|f(x)|^p\,dx=\int_{-\pi}^{\pi}[f(x)-T^*_{n-1}(x)]\,|f(x)|^{p-1}\operatorname{sign}f(x)\,dx$$

$$+\int_{-\pi}^{\pi}T^*_{n-1}(x)\,|f(x)|^{p-1}\operatorname{sign}f(x)\,dx\leqslant\int_{-\pi}^{\pi}|f(x)-T^*_{n-1}(x)|\,|f(x)|^{p-1}dx$$

$$+\frac{1}{2\pi}\int_{-\pi}^{\pi}T^*_{n-1}(x)\,dx\int_{-\pi}^{\pi}f(x)\,dx\leqslant E_n(f,L_p)\left(\int_{-\pi}^{\pi}|f(x)|^p\,dx\right)^{1-\frac{1}{p}}$$

$$+\frac{1}{2\pi}\overline{m}q\int_{-\pi}^{\pi}T^*_{n-1}(x)\,dx,$$

i.e.

$$\|f(x)\|_p\leqslant E_n(f,L_p)+\frac{\overline{m}q\,\|T^*_{n-1}(x)\|_p}{2\pi\left(\int_{-\pi}^{\pi}|f(x)|^p\,dx\right)^{1-\frac{1}{p}}}\leqslant E_n(f,L_p)+C\overline{m}q.$$

Since $\|f(x)\|_p=(4\pi\overline{m}-2\pi\overline{m}q)^{1/p}$, we have

$$E_n(f,L_p)\geqslant(4\pi\overline{m}-2\pi\overline{m}q)^{\frac{1}{p}}-C\overline{m}q.$$

Moreover, for arbitrary t $(0\leq t\leq 1)$ we have

$$\begin{aligned}\|f(x+2\pi t)-f(x)\|_p=\{2\pi\,[&\operatorname{mes}([0,1]-E_1\cup E_2)\cap(E_1\cup E_2)_t\\ &+\operatorname{mes}([0,1]-E_1\cup E_2)_t\cap(E_1\cup E_2)+\\ &+2^p(\operatorname{mes}E_1\cap(E_2)_t+\operatorname{mes}(E_1)_t\cap E_2)]\}^{\frac{1}{p}}\\ &\leqslant\{2\pi\,[4\overline{m}(1-2\overline{m})+2^{p+1}\overline{m}^2]+B\varepsilon\}^{\frac{1}{p}}\end{aligned}$$

by (22), where B is an absolute constant. Thus

$$\frac{E_n(f, L_p)}{\omega(t, f, L_p)} \geqslant \frac{(4\pi\bar{m} - 2\pi\bar{m}q)^{\frac{1}{p}} - C\bar{m}q}{(8\pi\bar{n} - 16\pi\bar{m}^2 + 2^{p+1}\bar{m}^2 + B\varepsilon)^{\frac{1}{p}}} \quad (0 < t \leqslant 2\pi).$$

Taking $\bar{m}$, ϵ and q sufficiently small, we get the conclusion of the theorem.

§2.

In this section we estimate

$$e(t, p) = \inf_{\|A(x)\|_1 = 1} \sup_{\|f(x)\|_p \leqslant 1} \frac{\|f(x) - Af(x)\|_p}{\omega(t, f, L_p)},$$

where $Af(x)$ is a linear operator of the form

$$Af(x) = \int_{-\pi}^{\pi} f(\tau) A(x - \tau)\, d\tau, \tag{24}$$

$$f(x) \in L_p \ (1 \leqslant p < \infty), \ A(x) \in L_1, \ 0 < t \leqslant 2\pi.$$

We shall need the following result of Fejér [**5**].

Lemma 4 (Fejér). *If $f(x) \in L_1$ and $g(x)$ has period 2π and is bounded, then*

$$\lim_{n \to \infty} \int_{-\pi}^{\pi} f(x) g(nx)\, dx = \frac{1}{2\pi} \int_{-\pi}^{\pi} f(x)\, dx \int_{-\pi}^{\pi} g(x)\, dx.$$

Let $S(x, h)$ be the Steklov kernel,

$$S(x, h) = \begin{cases} \dfrac{1}{h} & \left(-\dfrac{h}{2} \leqslant x \leqslant \dfrac{h}{2}\right), \\ 0 & \left(-\pi \leqslant x < -\dfrac{h}{2}, \ \dfrac{h}{2} < x \leqslant \pi\right), \end{cases} \quad (0 < h \leqslant 2\pi),$$

$$S(x, h) = S(x + 2\pi, h),$$

and let $f_h(x) = \int_{-\pi}^{\pi} f(\tau) S(x - \tau, h)\, d\tau$ be the Steklov function for $f(x)$.

Theorem 2. *Let $0 < t \leqq 2\pi$. We have*

$$\max\left\{2^{\frac{1-p}{p}}, 2^{-\frac{1}{p}}\right\} \leqslant e(t, p) \leqslant 1 \quad (1 \leqslant p < \infty). \tag{25}$$

Proof. It is known (see for example [**3**, 1965 ed., p. 225]) that

$$\|f(x) - f_{2t}(x)\|_p \leqslant \omega(t, f, L_p).$$

Moreover, $\|S(x, t)\|_1 = 1$, and hence $e(t, p) \leqq 1$.

Now let $A(x)$, with $\|A(x)\|_1 = 1$, be any operator of the form (24), and let $\epsilon > 0$ and $\alpha > 0$ be given. It was shown in Theorem 1 [cf. (20) and (21)] that

$$\omega\,(t,\ f\,(x,E,n),L_p) \leqslant 2\,(\pi+4\pi\varepsilon)^{\frac{1}{p}}.$$

for all t $(0 < t \leqq 2)$ and all n. By Fejér's Lemma we can take n so that

$$\left|\int_{-\pi}^{\pi} f(\tau,E,n)\,A\,(x-\tau)\,d\tau\right| < \alpha;$$

then

$$\|f_E(x) - Af_E(x)\|_p = (2\pi)^{\frac{1}{p}} + O(\alpha),$$

and hence

$$\frac{\|f_E(x) - Af_E(x)\|_p}{\omega\,(t,f_E,L_p)} \geqslant \frac{(2\pi)^{\frac{1}{p}} + O(\alpha)}{2\,(\pi+4\pi\varepsilon)^{\frac{1}{p}}} \quad (0 < t \leqslant 2\pi).$$

Reasoning similarly with $f(x) = f(x, E_1, E_2, n)$ [cf. (23)] we obtain

$$\overline{\lim_{n\to\infty}}\,\overline{\lim_{m\to 0}}\,\overline{\lim_{q\to 0}}\,\frac{\|f(x) - Af(x)\|_p}{\omega\,(t,f,L_p)} \geqslant 2^{-\frac{1}{p}} \quad (0 < t \leqslant 2\pi).$$

This completes the proof.

References

[1] D. Jackson, *Über die Genauigkeit der Annäherung stetiger Funktionen durch ganze rationale Funktionen gegebenen Grades und trigonometrische Summen gegebener Ordnung*, Dissertation, Göttingen, 1911.

[2] N. P. Korneĭčuk, *The exact constant in the theorem of D. Jackson on the best uniform approximation of continuous periodic functions*, Dokl. Akad. Nauk SSSR **145** (1962), 514-515 = Soviet Math. Dokl. **3** (1962), 1040-1041. MR **27** #521.

[3] N. I. Ahiezer, *Lectures on the theory of approximation*, 2nd rev. ed., "Nauka", Moscow, 1965; English transl. of 1st ed., Ungar, New York, 1956. MR **29** #1872; MR **32** #6108.

[4] S. M. Nikol'skiĭ, *Approximation of functions in the mean by trigonometric polynomials*, Izv. Akad. Nauk SSSR Ser. Mat. **10** (1946), 207-256. (Russian) MR **8**, 149.

[5] L. Fejér, *Lebesguesche Konstanten und divergent Fourier-reihen*, J. Reine. Angew. Math. **138** (1910), 22-53.

[6] N. I. Černyh, *On Jackson's inequality in L_2*, Trudy Mat. Inst. Steklov. **88** (1967), 71-74 = this issue, pp.75-78

A REMARK ON JACKSON'S THEOREM

S. B. STEČKIN

1. Let $f(x)$ be a function of period 2π, belonging to the space H, where H is C or L_p $(p \geqq 1)$. Put

$$E_n(f) = \min \| f - T_{n-1} \|,$$

where the minimum is taken over all trigonometric polynomials $T_{n-1}(x)$ of degree $n-1$, and

$$\omega(\delta, f) = \sup_{|h| \leqslant \delta} \| f(x+h) - f(x) \|.$$

By Jackson's theorem and its L_p analog there is an absolute constant K such that

$$E_n(f) \leqslant K\omega\left(\frac{1}{n}, f\right) \quad (n = 1, 2, \ldots). \tag{1}$$

It has recently become clear that it is better to write

$$E_n(f) \leqslant K\omega\left(\frac{\pi}{n}, f\right) \quad (n = 1, 2, \ldots). \tag{2}$$

For example, if we use the C metric it was shown by N. P. Korneĭčuk [1] that the best value of K in (2) is unity; and if we use the L_2 metric, it was shown by N. I. Černyh [2] that the best value of K is $1/\sqrt{2}$.

In addition to the best approximation $E_n(f)$ we consider the deviation of f from its approximation by a linear method $A_n(f)$ that maps our space into the space of trigonometric polynomials of degree $n-1$:

$$U_n(f) = \| f - A_n(f) \| \quad (n = 1, 2, \ldots).$$

In fact, Jackson showed that there is a linear method $A_n(f)$ such that

$$U_n(f) \leqslant K_1\omega\left(\frac{\pi}{n}, f\right) \tag{3}$$

for every $f \in C$.

Put

$$K^* = \sup_n \sup_f \frac{E_n(f)}{\omega\left(\frac{\pi}{n}, f\right)}, \quad K_1^* = \sup_n \inf_{A_n} \sup_f \frac{U_n(f)}{\omega\left(\frac{\pi}{n}, f\right)}. \tag{4}$$

It is clear that $K^* \leqq K_1^*$. Favard [4] showed that $K_1^* \leqq 3/2$ for C, and it follows from Černyh's theorem that $K^* = K_1^* = 1/\sqrt{2}$ for L_2. Since Korneĭčuk's approximation method is not linear, his result implies only that $1 = K^* \leqq K_1^*$ for C. The exact value of K_1^* is unknown for this case.

In the present note I show, by a well-known method ([4], pp. 245, 246), that $K_1^* \leqq 3/2$ in all cases; more precisely, *there is a linear approximation method* $A_n(f)$ *such that*

$$U_n(f) \leqslant \frac{3}{2}\omega\left(\frac{\pi}{n}, f\right) \quad (n = 1, 2, \ldots), \tag{5}$$

for every $f \in H$; *the approximation method is the same for all the spaces under consideration.*

2. We need a property of the Steklov function. For $f \in H$ put

$$S_h(f) = \frac{1}{h}\int_{-h/2}^{h/2} f(x+t)\,dt \quad (h > 0).$$

Then

$$\|f - S_h(f)\| = \left\|\frac{1}{h}\int_0^{h/2}\{f(x+t) - 2f(x) + f(x-t\}\,dt\right\|$$
$$\leqslant \frac{1}{h}\int_0^{h/2}\|f(x+t) - 2f(x) + f(x-t)\|\,dt \leqslant \frac{2}{h}\int_0^{h/2}\omega(t, f)\,dt \tag{6}$$

and (see [3], Chapter V)

$$\|S_h'(f)\| \leqslant \frac{1}{h}\omega(h, f). \tag{7}$$

We note that for the space C the estimate

$$\|f - S_h(f)\| \leqslant \frac{2}{h}\int_0^{h/2}\omega(t)\,dt \tag{8}$$

is best possible for any class $H[\omega]$ of functions $f(x)$ that satisfy $\omega(\delta, f) \leqq \omega(\delta)$ $(0 \leqq \delta \leqq \pi)$, where $\omega(\delta)$ is a given modulus of continuity (not necessarily concave). In fact, (8) becomes an equality for $f^*(x) = \omega(|x|)(|x| \leqq \pi)$, a function that, as is well known, belongs to $H[\omega]$.

3. By results of Favard, Ahiezer and Kreĭn (see for example [3, 1965 ed., pp. 250-251]), there is a linear approximation method $X_n(f)$ such that

$$\|f - X_n(f)\| \leqslant \frac{\pi}{2n}\|f'\| \tag{9}$$

for every $f \in H$ for which $f' \in H$.

Let us take as an approximation method

$$A_n(f) = X_n(S_h(f)). \tag{10}$$

Then we have

$$U_n(f) = \|f - A_n(f)\| \leqslant \|f - S_h(f)\| + \|S_h(f) - X_n(S_h(f))\|,$$

whence by (6), (9) and (7) we get

$$U_n(f) \leqslant \frac{2}{h}\int_0^{h/2}\omega(t,f)\,dt + \frac{\pi}{2n}\|S_h'(f)\| \leqslant \frac{2}{h}\int_0^{h/2}\omega(t,f)\,dt + \frac{\pi}{2nh}\omega(h,f). \tag{11}$$

Putting $h = 2\pi/n$ and using the monotonic character of the modulus of continuity together with the estimate $\omega(h,f) \leqq 2\omega(h/2,f)$, we infer that

$$U_n(f) \leqslant \omega\left(\frac{h}{2},f\right) + \frac{\pi}{nh}\omega\left(\frac{h}{2},f\right) = \frac{3}{2}\omega\left(\frac{\pi}{n},f\right) \quad (n = 1, 2, \ldots),$$

and (5) is established.

Taking $h = \pi/n$, we obtain similarly

$$U_n(f) \leqslant 2\omega\left(\frac{\pi}{2n},f\right) \quad (n = 1, 2, \ldots). \tag{12}$$

Now consider concave moduli of continuity. If $\omega(\delta,f) \leqq \omega(\delta)$, where $\omega(\delta)$ is a concave function, then by applying Jensen's theorem twice to (11) and putting $h = (\pi/n)(1+\sqrt{3})$ we obtain

$$\begin{aligned} U_n(f) &\leqslant \omega\left(\frac{h}{4}\right) + \frac{\pi}{2nh}\omega(h) \leqslant \left(1 + \frac{\pi}{2nh}\right)\omega\left(\frac{\frac{h}{4} + \frac{\pi}{2n}}{1 + \frac{\pi}{2nh}}\right) \\ &= \frac{3+\sqrt{3}}{4}\omega\left(\frac{\pi}{n}\right) \qquad (n = 1, 2, \ldots). \end{aligned} \tag{13}$$

We note that (5), (12) and (13) remain true for any homogeneous space of continuous functions, i.e. for a space satisfying the following conditions:

1) $\|f(x+t)\| = \|f(x)\|$ for all t;

2) $\|f(x+h) - f(x)\| \to 0$ $(h \to 0)$ for each element f of the space.

Inequalities (6), (7) and (9) remain true for such spaces; the proofs are unchanged.

References

[1] N. P. Korneĭčuk, *The exact constant in the theorem of D. Jackson on the best uniform approximation of continuous periodic functions,* Dokl. Akad. Nauk SSSR **145** (1962), 514-515 = Soviet Math. Dokl. **3** (1962), 1040-1041. MR **27** # 521.

[2] N. I. Černyh, *On Jackson's inequality in* L_2, Trudy Mat. Inst. Steklov. **88** (1967), 71-74 = this issue, pp. 75-78

[3] N. I. Ahiezer, *Lectures on the theory of approximation,* 2nd rev. ed., "Nauka", Moscow, 1965; English transl. of 1st ed., Ungar, New York, 1956. MR **20** # 1872; MR **32** # 6108.

[4] J. Favard, *Sur les meilleurs procédés d'approximation de certaines classes de fonctions par des polynomes trigonométriques,* Bull. Sci. Math. (2) **61** (1937), 209-224, 243-256.

THE APPROXIMATION OF DIFFERENTIABLE FUNCTIONS BY TRIGONOMETRIC POLYNOMIALS IN THE L METRIC

S. B. STEČKIN AND S. A. TELJAKOVSKIĬ

§1.

There are many papers dealing with the uniform approximation of periodic differentiable functions by various mean values formed from their Fourier series. The analogous problem of the approximation of functions in the L metric has been studied much less. The most fundamental results in this direction were obtained by S. M. Nikol'skiĭ [1]. He showed that for many methods of approximation the upper bounds for the deviation in the L and C metrics are, for various classes of differentiable functions, asymptotically equal or even the same.

In the present paper we show that all results on upper bounds for uniform approximation of periodic differentiable functions of the classes[1] W_α^r by mean values formed from their Fourier series can be carried over to the L metric except for supplementary terms which in many cases decrease more rapidly than the principal terms.

In §2 we give some general propositions about the norms of operators of convolution type.

In §3 we discuss the approximation of periodic functions.

In §4 we construct a method of approximation for which the upper bounds of the deviations in the C and L metrics are not asymptotically equal for the class W_0^1. In this case the supplementary terms are of the same order as the principal term.

§2.

Let M be the space of bounded measurable functions defined on $[-\pi, \pi]$, with norm the essential upper bound,

$$\|\varphi(x)\|_M = \sup_x \operatorname{vrai} |\varphi(x)|,$$

and let M_0 be the subspace of M consisting of functions with mean value zero.

If $K(x)$ is a summable function of period 2π, the convolution

$$U(\varphi) = \int_{-\pi}^{\pi} K(t-x)\varphi(t)\,dt \tag{1}$$

[1] These classes are defined in §3.

is a linear operator from M to C and from M_0 to C_0, where C and C_0 are the subspaces of continuous functions belonging to M and M_0.

We denote the norms of these operators by $\|U\|_M$ and $\|U\|_{M_0}$. As is well known,

$$\|U\|_M = \int_{-\pi}^{\pi} |K(t)|\,dt. \tag{2}$$

In addition, B. Sz.-Nagy [2] (cf. also [1]) showed that

$$\|U\|_{M_0} = \min_a \int_{-\pi}^{\pi} |K(t) - a|\,dt = \int_{-\pi}^{\pi} |K(t) - a^*|\,dt. \tag{3}$$

Moreover, M contains functions for which the norm (2) is attained, and M_0 contains functions for which the norm (3) is attained. In both cases the extremal functions depend on the kernel $K(t)$.

In addition, let V be the space of functions of bounded variation defined on $[-\pi, \pi]$, with norm the total variation,

$$\|\varphi(x)\|_V = \int |d\varphi(x)|.$$

The subspace of periodic functions in V, i.e. those for which

$$\int_{-\pi}^{\pi} d\varphi(x) = 0,$$

will be denoted by V_0.

If, as before, $K(x)$ is a periodic summable function, the convolution[2]

$$U^*(\varphi) = \int_{-\pi}^{\pi} K(t - x)\,d\varphi(t) \tag{4}$$

is a linear operator from V to the space L of summable functions, or from V_0 to L_0, where L_0 is the subspace of L consisting of functions with mean value zero. We denote the norms of these operators by $\|U^*\|_V$ and $\|U^*\|_{V_0}$. It is known that

$$\|U^*\|_V = \int_{-\pi}^{\pi} |K(t)|\,dt, \tag{5}$$

and the norm is attained for

$$\chi(t) = \begin{cases} 0 & \text{for } -\pi \leqslant t \leqslant 0, \\ 1 & \text{for } 0 < t \leqslant \pi, \end{cases} \tag{6}$$

which is independent of $K(t)$.

---

[2] Since $K(x)$ is summable, $K(t - x)$, as a function of two variables, is $|x\varphi(t)|$-measurable, and the integral (4) exists for almost all x and is summable. In the cases considered in §3 the existence of this integral is evident.

It follows from (2) and (5) that

$$\|U\|_M = \|U^*\|_V. \tag{7}$$

Finally,

$$\|U^*\|_{V_0} = \frac{1}{2}\max_u \int_{-\pi}^{\pi} |K(t) - K(t+u)|\,dt. \tag{8}$$

Nikol'skiĭ [1, §2, Theorem 3] showed that the right side of (8) is the norm of (1) as an operator from L_0 to L_0. We shall not prove (8) here, since the proof is a duplication of Nikol'skiĭ's.

However, although there is not always a function in L_0 for which the norm of (1) is attained, to each kernel $K(t)$ there corresponds a function in V_0 for which $\|U^*(\varphi)\|_L = \|U^*\|_{V_0}$. In fact, there is equality for the function $\frac{1}{2}[\chi(t) - \chi(t+u_0)]$, where u_0 is the value of u for which the integral on the right of (8) attains its maximum. Hence in this case the extremal function depends on $K(t)$.

Comparison of (3) and (8) shows that (cf. [1])

$$\|U^*\|_{V_0} \leqslant \|U\|_{M_0}. \tag{9}$$

In addition, it follows from Nikol'skiĭ's results [1] that the equality

$$\|U^*\|_{V_0} = \|U\|_{M_0} \tag{10}$$

holds if and only if there is a u_0 such that

$$K^*(t)\,K^*(t+u_0) \leqslant 0 \tag{11}$$

for almost all t; here $K^*(t) = K(t) - a^*$, where a^* is defined in (3).

We shall obtain an approximate formula for $\|U\|_{M_0}$. Clearly

$$\|U\|_{M_0} \leqslant \|U\|_M = \int_{-\pi}^{\pi} |K(t)|\,dt.$$

On the other hand, put

$$\varphi_0(t) = \operatorname{sign} K(t) \quad \text{for } |t| < \frac{\pi}{2}, \qquad \varphi_0(t+\pi) = -\varphi_0(t).$$

Then $\varphi_0(t) \in M_0$, and consequently

$$\|U\|_{M_0} \geqslant \int_{-\pi}^{\pi} K(t)\,\varphi_0(t)\,dt \geqslant \int_{-\frac{\pi}{2}}^{\frac{\pi}{2}} |K(t)|\,dt - \int_{\frac{\pi}{2}\leqslant|t|\leqslant\pi} |K(t)|\,dt$$

$$= \int_{-\pi}^{\pi} |K(t)|\,dt - 2\int_{\frac{\pi}{2}\leqslant|t|\leqslant\pi} |K(t)|\,dt.$$

Hence

$$\|U\|_{M_0} = \int_{-\pi}^{\pi} |K(t)|\,dt - 2\theta_1 \int_{\frac{\pi}{2}\leqslant|t|\leqslant\pi} |K(t)|\,dt \quad (0\leqslant\theta_1\leqslant 1). \tag{12}$$

We shall show that there is a similar formula for $\|U^*\|_{V_0}$. We have

$$\|U^*\|_{V_0} \leqslant \|U^*\|_V = \int_{-\pi}^{\pi} |K(t)|\,dt$$

and by (8)

$$\|U^*\|_{V_0} \geqslant \frac{1}{2}\int_{-\pi}^{\pi} |K(t) - K(t+\pi)|\,dt$$

$$\geqslant \int_{-\pi}^{\pi} |K(t)|\,dt - 2 \int_{\frac{\pi}{2}\leqslant|t|\leqslant\pi} |K(t)|\,dt,$$

whence

$$\|U^*\|_{V_0} = \int_{-\pi}^{\pi} |K(t)|\,dt - 2\theta_2 \int_{\frac{\pi}{2}\leqslant|t|\leqslant\pi} |K(t)|\,dt \quad (0\leqslant\theta_2\leqslant 1). \tag{13}$$

Note that for $\varphi_1(t) = \frac{1}{2}[\chi(t) - \chi(t+\pi)]$, which is independent of $K(t)$, we have

$$\|U^*(\varphi_1)\|_L = \frac{1}{2}\int_{-\pi}^{\pi} |K(t) - K(t+\pi)|\,dt$$

$$= \int_{-\pi}^{\pi} |K(t)|\,dt - 2\theta_3 \int_{\frac{\pi}{2}\leqslant|t|\leqslant\pi} |K(t)|\,dt \quad (0\leqslant\theta_3\leqslant 1),$$

where θ_3 in general depends on $K(t)$. Combining (12) and (13) and using (9), we obtain the following result.

THEOREM. *For any summable $K(t)$ we have*

$$\|U^*\|_{V_0} = \|U\|_{M_0} - 2\theta \int_{\frac{\pi}{2}\leqslant|t|\leqslant\pi} |K(t)|\,dt \quad (0\leqslant\theta\leqslant 1). \tag{14}$$

§3.

We shall consider the approximation of differentiable functions by linear means formed from their Fourier series.

Let $r > 0$, let α be a real number, and put

$$K_\alpha^r(x) = \sum_{k=1}^{\infty} \frac{1}{k^r}\cos\left(kx - \frac{\alpha\pi}{2}\right).$$

The class of functions $f(x)$ representable as convolutions

$$f(x)=\frac{1}{\pi}\int_{-\pi}^{\pi} K_\alpha^r(t-x)\varphi(t)\,dt, \tag{15}$$

where $\varphi(t)\in M_0$ and $\|\varphi(t)\|_M\leqq 1$, is denoted by $W_\alpha^r C$ (cf. [**2,3**]).

The class of functions $f(x)$ representable in the form

$$f(x)=\frac{1}{\pi}\int_{-\pi}^{\pi} K_\alpha^r(t-x)\,d\varphi(t), \tag{16}$$

where $\varphi(t)\in V_0$ and $\|\varphi(t)\|_V\leqq 1$, is denoted by $Q_\alpha^r L$ (cf. [**10**, Appendix, 101]).

It is clear that an $f(x)$ in one of these classes is necessarily summable. Let

$$\frac{a_0}{2}+\sum_{k=1}^{\infty}(a_k\cos kx+b_k\sin kx) \tag{17}$$

be its Fourier series. Then the class $W_\alpha^r C$ is the class of functions for which

$$\sum_{k=1}^{\infty} k^r\left[a_k\cos\left(kx+\frac{\alpha\pi}{2}\right)+b_k\sin\left(kx+\frac{\alpha\pi}{2}\right)\right] \tag{18}$$

is the Fourier series of a function φ as in (15), and $W_\alpha^r L$ is the class of functions for which (18) is the Fourier series of $d\varphi$ as in (16).

For $\alpha=r$ the series (18) is the Fourier series of the derivative of $f(x)$ of order r in the sense of Weyl; we then call the classes $W^r C$ and $W^r L$. For $\alpha=r-1$ we obtain the classes of functions conjugate to functions of classes $W^r C$ and $W^r L$; we denote them by $\overline{W}^r C$ and $\overline{W}^r L$.

By using the matrix $\{\lambda_{n,k}\}$, $n,k=1,2,\cdots$; $\lambda_{n,k}=0$ for $k\geqq n$, we associate with each summable $f(x)$ with Fourier series (17) a polynomial of degree $n-1$:

$$u_n(f,x)=\frac{a_0}{2}+\sum_{k=1}^{n-1}\lambda_{n,k}(a_k\cos kx+b_k\sin kx).$$

We consider the upper bound of the deviation of $u_n(f,x)$ from $f(x)$ when $f(x)$ belongs to $W_\alpha^r C$ or $W_\alpha^r L$:

$$U_n(W_\alpha^r,C)=\sup_{f\in W_\alpha^r C}\|f(x)-u_n(f,x)\|_C \tag{19}$$

and

$$U_n(W_\alpha^r,L)=\sup_{f\in W_\alpha^r L}\|f(x)-u_n(f,x)\|_L. \tag{20}$$

In investigating the asymptotic behavior of $U_n(W_\alpha^r,C)$ as $n\to\infty$ it is often convenient to represent the difference $f(x)-u_n(f,x)$ as a convolution with a kernel defined on the whole real axis (cf. [**4,5**]). We shall obtain a similar formula for $U_n(W_\alpha^r,L)$.

Let $\{\mu_n(u)\}$ be a sequence of continuous functions defined for $u\geqq 0$, with

$$\mu_n\left(\frac{k}{n}\right)=(1-\lambda_{n,k})\left(\frac{k}{n}\right)^{-r};\ n, k=1, 2, \ldots \tag{21}$$

Suppose also that the Fourier transform of $\mu_n(u)$,

$$\frac{1}{\pi}\int_0^\infty \mu_n(u)\cos\left(ut+\frac{\alpha\pi}{2}\right)du,$$

is summable on $(-\infty, \infty)$, i.e. that

$$A(\mu_n)=\frac{1}{\pi}\int_{-\infty}^{\infty}\left|\int_0^\infty \mu_n(u)\cos\left(ut+\frac{\alpha\pi}{2}\right)du\right|dt \tag{22}$$

converges.

It was shown in [5] that under our hypotheses on $\mu_n(u)$ we have

$$\mu_n(x)=\frac{1}{\pi}\int_{-\infty}^{\infty}\cos\left(xt+\frac{\alpha\pi}{2}\right)\int^{\infty}\mu_n(u)\cos\left(ut+\frac{\alpha\pi}{2}\right)du\,dt \tag{23}$$

for all $x \geq 0$.

Let $f(x) \in W_\alpha^r L$ and let $f_{\alpha-1}^{r-1}(x)$ be an element of V for which (16) holds. It is easily verified that

$$\frac{1}{\pi n^r}\int_{-\infty}^{\infty}\left[\int_0^\infty \mu_n\left(\frac{u}{n}\right)\cos\left(u(t-x)+\frac{\alpha\pi}{2}\right)du\right]df_{\alpha-1}^{r-1}(t), \tag{24}$$

considered as an improper Lebesgue-Stieltjes integral, converges for almost all x and represents a periodic summable function. By using (23) it can be seen that the Fourier series of (24) is

$$\sum_{k=1}^{\infty}\mu_n\left(\frac{k}{n}\right)\left(\frac{k}{n}\right)^r(a_k\cos kx+b_k\sin kx). \tag{25}$$

But since $\mu_n(k/n)$ satisfies (21), the series (25) is the Fourier series of $f(x)-u_n(f,x)$. Therefore

$$f(x)-u_n(f,x)=\frac{1}{\pi n^r}\int_{-\infty}^{\infty}\int_0^\infty \mu_n\left(\frac{u}{n}\right)\cos\left(u(t-x)+\frac{\alpha\pi}{2}\right)du\,df_{\alpha-1}^{r-1}(t) \tag{26}$$

almost everywhere.

It follows from (26) that

$$\begin{aligned}\|f(x)-u_n(f,x)\|_L=\frac{1}{\pi n^r}\left\|\int_{-\pi}^{\pi}\int_0^\infty \mu_n\left(\frac{u}{n}\right)\cos\left(u(t-x)+\frac{\alpha\pi}{2}\right)du\,df_{\alpha-1}^{r-1}(t)\right\|_L\\ +O\left(\frac{1}{n^r}\int_{\pi\leqslant|t|}\left|\int_0^\infty \mu_n\left(\frac{u}{n}\right)\cos\left(ut+\frac{\alpha\pi}{2}\right)du\right|dt\right),\end{aligned} \tag{27}$$

uniformly for $f(x) \in W_\alpha^r L$. Hence, according to (13), we obtain

$$U_n(W^r_\alpha, L) = \frac{1}{\pi n^r} \int_{-\pi}^{\pi} \left| \int_0^\infty \mu_n\left(\frac{u}{n}\right) \cos\left(ut + \frac{\alpha\pi}{2}\right) du \right| dt$$

$$+ O\left(\frac{1}{n^r} \int_{\frac{\pi}{2} \leqslant |t|} \left| \int_0^\infty \mu_n\left(\frac{u}{n}\right) \cos\left(ut + \frac{\alpha\pi}{2}\right) du \right| dt\right) \tag{28}$$

$$= \frac{1}{n^r} A(\mu_n) + O\left(\frac{1}{n^r} \int_{\frac{\pi n}{2} \leqslant |t|} \left| \int_0^\infty \mu_n(u) \cos\left(ut + \frac{\alpha\pi}{2}\right) du \right| dt\right).$$

Since (26) holds for $f(x) \in W^r_\alpha L$, it also holds for $f(x) \in W^r_\alpha C$. By a similar argument we obtain from (26)

$$U_n(W^r_\alpha, C) = \frac{1}{n^r} A(\mu_n) + O\left(\frac{1}{n^r} \int_{\frac{\pi n}{2} \leqslant |t|} \left| \int_0^\infty \mu_n(u) \cos\left(ut + \frac{\alpha\pi}{2}\right) du \right| dt\right), \tag{29}$$

i.e. the upper bounds $U_n(W^r_\alpha, C)$ and $U_n(W^r_\alpha, L)$ differ only by an amount whose order does not exceed

$$\frac{1}{n^r} \int_{\frac{\pi n}{2} \leqslant |t|} \left| \int_0^\infty \mu_n(u) \cos\left(ut + \frac{\alpha\pi}{2}\right) du \right| dt.$$

Formula (29) was established in [5].

We note that the extremal function in (29) for the space C depends on the summability method, whereas for the space L the extremal function in (28) can be chosen independently of the summability method, as the remark after (13) shows.

We see that the asymptotic formula for $U_n(W^r_\alpha, C)$ that follows from (29) will be valid for approximation in the L metric also. Hence the results of [5] on the upper bound of the deviation can be extended to the L metric. Equally, the results of [6] can be extended to approximation in the L metric. In particular, we obtain, for example, the result of I. E. Gopengauz and A. L. Rabinovič [7].

§4.

We now construct an example of an approximation method for which $U_n(W^r_\alpha, C)$ and $U_n(W^r_\alpha, L)$ are not asymptotically equal as $n \to \infty$.

We consider the Vallée Poussin sums

$$v_{n,m}(f, x) = \frac{1}{m} \sum_{k=n-m}^{n-1} s_k(f, x); \quad m = 1, 2, \ldots, n; \quad n = 1, 2, \ldots,$$

where $s_k(f, x)$ are the partial sums of order k of the Fourier series of $f(x)$. For $m = n$ we obtain the Fejér sums $\sigma_n(f, x)$. We denote the upper bounds (19) and (20) for the Vallée Poussin sums by $V_{n,m}(W^r_\alpha, C)$ and $V_{n,m}(W^r_\alpha, L)$.

We construct our example for $m = n - 2$ and the classes $\overline{W}^1C$ and $\overline{W}^1L$.

It was remarked in [5] that if $n - m = p$ is fixed, $p \geqq 1$, then both terms in (29) decrease at the same rate as $n \to \infty$. The following asymptotic formula for this case was established in [8]:

$$V_{n,m}(\overline{W}^1, C) = \frac{n}{m} V_{n,n}(\overline{W}^1, C) + \frac{p}{m} V_{p,p}(\overline{W}^1, C) + O\left(\frac{1}{n}\sqrt{\frac{\log n}{n}}\right). \quad (30)$$

From this and Stečkin's [9] asymptotic formula for the approximate Fejér sums,

$$V_{n,n}(\overline{W}^1, C) = \frac{2}{\pi n}\int_0^\infty \left|\int_y^\infty \frac{\sin t}{t^2}\,dt\right| dy + O\left(\frac{1}{n^2}\right), \quad (31)$$

we obtain

$$V_{n,m}(\overline{W}^1, C) = \left\{\frac{2}{\pi}\int_0^\infty \left|\int_y^\infty \frac{\sin t}{t^2}\,dt\right| dy + pV_{p,p}(\overline{W}^1, C)\right\}\frac{1}{n} + O\left(\frac{1}{n}\sqrt{\frac{\log n}{n}}\right). \quad (32)$$

We shall establish a formula similar to (30) for $V_{n,m}(\overline{W}^1, L)$. Since

$$f(x) - v_{n,m}(f, x) = \frac{n}{m}[f(x) - \sigma_n(f, x)] - \frac{p}{m}[f(x) - \sigma_p(f, x)], \quad (33)$$

we have

$$V_{n,m}(\overline{W}^1, L) \leqslant \frac{n}{m} V_{n,n}(\overline{W}^1, L) + \frac{p}{m} V_{p,p}(\overline{W}^1, L). \quad (34)$$

To estimate $V_{n,m}(\overline{W}^1, L)$ from below we need a formula for the exact value of the upper bound $V_{n,m}(\overline{W}^1, L)$.

If $f(x) \in \overline{W}^1L$, then for almost all x (in any case for the values of x for which $f'(x)$ exists)

$$\begin{aligned} f(x) - \sigma_n(f, x) &= \lim_{\varepsilon \to +0} \frac{1}{\pi n}\int_\varepsilon^{2\pi-\varepsilon} \bar{f}(x+t)\frac{\sin nt}{4\sin^2\frac{t}{2}}\,dt \\ &= \lim_{\varepsilon \to 0}\frac{1}{\pi n}\left\{\bar{f}(x+y)\int_\pi^y \frac{\sin nt}{4\sin^2\frac{t}{2}}\,dt\,\Bigg|_{y=\varepsilon}^{2\pi-\varepsilon}\right. \\ &\left. - \int_\varepsilon^{2\pi-\varepsilon}\int_\pi^y \frac{\sin nt}{4\sin^2\frac{t}{2}}\,dt\,d_y\bar{f}(x+y)\right\} = \frac{1}{\pi n}\int_0^{2\pi}\int_y^\pi \frac{\sin nt}{4\sin^2\frac{t}{2}}\,dt\,d_y\bar{f}(x+y). \end{aligned} \quad (35)$$

Let $F_n(y)$ be the periodic function which is

$$\int_y \frac{\sin nt}{4\sin^2\frac{t}{2}}\,dt$$

on $[0, 2\pi]$. From (8) we have

$$\begin{aligned} V_{n,n}(\overline{W}^1, L) &= \frac{1}{2\pi n} \max_u \int_0^{2\pi} | F_n(y) - F_n(y+u) | \, dy \\ &= \frac{1}{2\pi n} \int_0^{2\pi} | F_n(y) - F_n(y+u_n) | \, dy, \end{aligned} \tag{36}$$

where u_n, $0 < u_n < 2\pi$, is a number for which the maximum is attained.

Using (33), (35) and (8), we obtain

$$V_{n,m}(\overline{W}^1, L) = \frac{1}{2\pi m} \max_u \int_0^{2\pi} | F_n(y) - F_n(y+u) - F_p(y) + F_p(y+u) | \, du. \tag{37}$$

If n is sufficiently large,

$$b = b_n = \frac{1}{\sqrt{n \log n}} < \frac{1}{2} \min (u_p, 2\pi - u_p).$$

Introduce the sets $E_1 = [0, b] \cup [2\pi - b, 2\pi]$, $E_2 = [2\pi - u_p - b, 2\pi - u_p + b]$ and $E_3 = [0, 2\pi] \backslash (E_1 \cup E_2)$. We see from (37) that as $n \to \infty$

$$\begin{aligned} V_{n,m}(\overline{W}^1, L) &\geqslant \frac{1}{2\pi m} \int_0^{2\pi} | F_n(y) - F_n(y+u_p) - F_p(y) + F_p(y+u_p) | \, dy \\ &= \frac{1}{2\pi m} \Big\{ \int_{E_1} | F_n(y) | \, dy + \int_{E_2} | F_n(y+u_p) | \, dy \\ &\quad + \int_{E_3} | F_p(y) - F_p(y+u_p) | \, dy \Big\} + O\Big(\frac{1}{m} \int_{E_2 \cup E_3} | F_n(y) | \, dy \\ &\quad + \frac{1}{m} \int_{E_1 \cup E_3} | F_n(y+u_p) | \, dy + \frac{1}{m} \int_{E_1 \cup E_2} | F_p(y) - F_p(y+u_p) | \, dy \Big) \\ &= \frac{1}{2\pi m} \Big\{ 2 \int_0^{2\pi} | F_n(y) | \, dy + \int_0^{2\pi} | F_p(y) - F_p(y+u_p) | \, dy \Big\} \\ &\quad + O\Big(\frac{1}{m} \int_{E_2 \cup E_3} | F_n(y) | \, dy + \frac{1}{m} \int_{E_1 \cup E_2} | F_p(y) - F_p(y+u_p) | \, dy \Big). \end{aligned}$$

From the estimates given in [8, p. 239] for the last two integrals, it follows that

$$\begin{aligned} &\frac{1}{m} \int_{E_2 \cup E_3} | F_n(y) | \, dy + \frac{1}{m} \int_{E_1 \cup E_2} | F_p(y) - F_p(y+u_p) | \, dy \\ &\qquad = O\Big(\frac{1}{mnb} + \frac{b}{m} \log \frac{1}{b} \Big) = O\Big(\frac{1}{n} \sqrt{\frac{\log n}{n}} \Big). \end{aligned}$$

From these estimates and (13) and (36) we obtain

$$V_{n,m}(\overline{W}^1, L) \geqslant \frac{n}{m} V_{n,n}(\overline{W}^1, L) + \frac{p}{m} V_{p,p}(\overline{W}^1, L) + O\left(\frac{1}{n}\sqrt{\frac{\log n}{n}}\right), \quad n \to \infty.$$

From this and (34) we obtain the asymptotic formula

$$V_{n,m}(\overline{W}^1, L) = \frac{n}{m} V_{n,n}(\overline{W}^1, L) + \frac{p}{m} V_{p,p}(\overline{W}^1, L) + O\left(\frac{1}{n}\sqrt{\frac{\log n}{n}}\right). \quad (38)$$

Equation (31) was deduced from (29) in [5]; hence it also holds for $V_{n,m}(\overline{W}^1, L)$. Hence for a fixed $p \geqq 1$ we have an asymptotic formula for $V_{n,m}(\overline{W}^1, L)$ corresponding to (32):

$$V_{n,m}(\overline{W}^1, L) = \left\{\frac{2}{\pi}\int_0^\infty \left|\int_y^\infty \frac{\sin t}{t^2}\,dt\right| dy + pV_{p,p}(\overline{W}^1, L)\right\}\frac{1}{n} + O\left(\frac{1}{n}\sqrt{\frac{\log n}{n}}\right). \quad (39)$$

It remains to show that $V_{p,,}(\overline{W}^1, C)$ and $V_{p,p}(\overline{W}^1, L)$ are not necessarily equal.

It is easily shown that $V_{1,1}(\overline{W}^1, C) = V_{1,1}(\overline{W}^1, L)$. However, even for $p = 2$ we have

$$V_{2,2}(\overline{W}^1, C) > V_{2,2}(\overline{W}^1, L). \quad (40)$$

In fact by (3)

$$V_{2,2}(\overline{W}^1, C) = \frac{1}{2\pi}\min_a \int_0^{2\pi} |F_2(y) - a|\,dy = \frac{1}{2\pi}\int_0^{2\pi} |F_2(y) - a_2|\,dy, \quad (41)$$

where a_2 minimizes the integral. Since $F_2(y) - a$ does not vanish on a set of positive measure for any a, the number a_2 has the property that (cf. [1, 2])

$$\int_0^{2\pi} \operatorname{sign}(F_2(y) - a_2)\,dy = 0.$$

It follows that $a_2 > 0$ (see the figure).

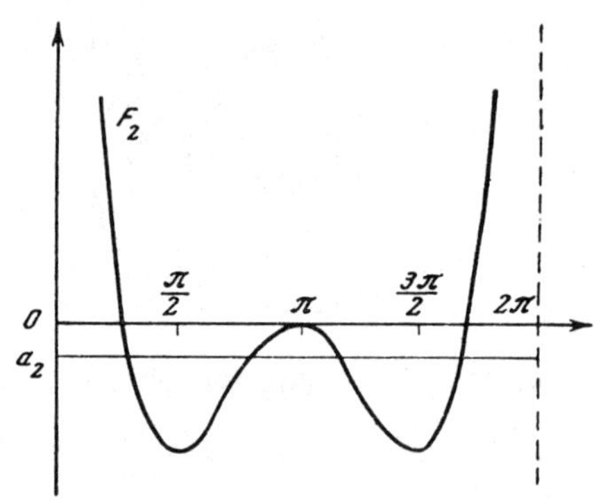

Put $F_2(y) - a_2 = F_2^*(y)$. By (8),

$$\begin{aligned} V_{2,2}(\overline{W}^1, L) &= \frac{1}{4\pi} \max_u \int_0^{2\pi} |F_2(y) - F_2(y+u)|\, dy \\ &= \frac{1}{4\pi} \max_u \int_0^{2\pi} |F_2^*(y) - F_2^*(y+u)|\, dy. \end{aligned} \tag{42}$$

But the upper bounds (41) and (42) are the same if and only if (cf. §2) there is a u for which

$$F_2^*(y+u)\, F_2^*(y) \leqslant 0 \tag{43}$$

almost everywhere.

For (43) to hold it is necessary that $F_2^*(y)$ changes sign at the points $\pi/4$, $3\pi/4$, $5\pi/4$, $7\pi/4$. But

$$F_2\left(\frac{\pi}{4}\right) = \int_{\frac{\pi}{4}}^{\pi} \frac{\sin 2t}{4\sin^2 \frac{t}{2}}\, dt = -1 - \frac{1}{\sqrt{2}} - 2 \log \sin \frac{\pi}{8} > 0,$$

and so certainly $F_2^*(\pi/4) > 0$.

Hence (40) follows. We have therefore established that $V_{n,n-2}(\overline{W}^1, C)$ and $V_{n,n-2}(\overline{W}^1, L)$ are not asymptotically equal as $n \to \infty$.

References

[1] S. M. Nikol'skiĭ, *Approximation of functions in the mean by trigonometric polynomials*, Izv. Akad. Nauk SSSR Ser. Mat. **10** (1946), 207-256. (Russian) MR **8**, 149.

[2] B. Sz.-Nagy, *Über gewisse Extremalfragen bei transformierten trigonometrischen Entwicklungen. I. Periodischer Fall*, Ber. Math.-Phys. Kl. Akad. Wiss. Leipzig **90** (1938), 103-134.

[3] S. B. Stečkin, *On the best approximation of certain classes of periodic functions by trigonometric polynomials*, Izv. Akad. Nauk SSSR Ser. Mat. **20** (1956), 643-648. (Russian) MR **18**, 303.

[4] B. Sz.-Nagy, *Sur une classe générale de procédés de sommation pour les séries de Fourier*, Hungarica Acta Math. **1** (1948), no. 3, 14-52. MR **10**, 369.

[5] S. A. Teljakovskiĭ, *On norms of trigonometric polynomials and approximation of differentiable functions by linear averages of their Fourier series*. I, Trudy Mat. Inst. Steklov. **62** (1961), 61-97; English transl., Amer. Math. Soc. Transl. (2) **28** (1963), 283-322. MR **27** #2777a; MR **27** #3995.

[6] ———, *On norms of trigonometric polynomials and approximation of differentiable functions by linear averages of their Fourier series*. II, Izv. Akad. Nauk SSSR Ser. Mat. **27** (1963), 253-272. (Russian) MR **27** #2777b.

[7] I. E. Gopengauz and A. L. Rabinovič, *A relation in the theory of uniform and mean approximation*, Ukrain. Mat. Ž. **12** (1960), 339-341. (Russian) MR **27** #1761.

[8] S. A. Teljakovskii, *Approximations to differentiable functions by linear means of their Fourier series*, Izv. Akad. Nauk SSSR Ser. Mat. **24** (1960), 213-242. (Russian) MR **22** #3937.

[9] ———, *Approximation of differentiable functions by de la Vallée Poussin's sums*, Dokl. Akad. Nauk SSSR **121** (1958), 426-429. (Russian) MR **20** #6632.

[10] N. I. Ahiezer, *Lectures on the theory of approximation*, 2nd rev. ed., "Nauka", Moscow, 1965; English transl., Ungar, New York, 1956. MR **20** #1872; MR **32** #6108.

INTERPOLATION BY FUNCTIONS WITH nth DERIVATIVE OF MINIMUM NORM

JU. N. SUBBOTIN

In this paper we consider the following problem of interpolation by functions. Let n be a fixed positive integer and let the values $y_m = f(m)$ be given at the integral points $m = 0, \pm 1, \pm 2, \cdots$; let the nth differences

$$\Delta^n y_m = \sum_{k=0}^{n} (-1)^{n-k} C_n^k y_{m+k}$$

satisfy

$$\sum_{m=-\infty}^{\infty} |\Delta^n y_m|^p \leqslant 1 \quad (1 \leqslant p < \infty); \tag{1}$$

we are to find the smallest value, for all sequences $\{y_n\}$ satisfying (1), of the constant $A_{n,p}$ in the inequality

$$\inf \|f^{(n)}(x)\|_{L_p} = \inf \Big(\int_{-\infty}^{\infty} |f^{(n)}(x)|^p \, dx \Big)^{\frac{1}{p}} \leqslant A_{n,p} \quad (1 \leqslant p < \infty),$$

where inf is taken over all functions for which $f(k) = y_k$ $(k = 0, \pm 1, \pm 2, \cdots)$.

The analogous problem for $p = \infty$ was considered in [1]. In this case (1) becomes

$$|\Delta^n y_m| \leqslant 1 \qquad (m = 0, \pm 1, \pm 2, \ldots)$$

and $\|f^{(n)}(x)\|_{L_\infty} = \sup_x |f^{(n)}(x)|$. We quote the results that we shall need from this paper.

THEOREM 1. *If the polynomial*

$$P_{2n}(x) = \sum_{l=0}^{2n} b_l x^l$$

satisfies

$$P_{2n}(x) \equiv x^{2n} P_{2n}\left(\frac{1}{x}\right),$$

i.e. is symmetric, and its zeros are distinct and not on the unit circumference, and if C_m is a bounded sequence, then the difference equation

$$\sum_{l=0}^{2n} b_l Z_{m+l} = C_m \quad (m = 0, \pm 1, \pm 2, \ldots)$$

has a unique bounded solution; this solution can be represented in the form

$$Z_m^0 = \sum_{k=-\infty}^{\infty} \sum_{p=1}^{n} \frac{x_p^{n-1+|k|}}{P'_{2n}(x_p)} C_{m-n+k}, \tag{2}$$

where x_p $(p = 1, 2, \cdots, n)$ are the zeros of $P_{2n}(x)$ inside $|x| < 1$. If, in addition, these zeros are all negative, then

$$\sup_m |Z_m^0| \leqslant \frac{\sup_m |C_m|}{|P_{2n}(-1)|}. \tag{3}$$

LEMMA A. *The polynomial*

$$Q_n(x) = (-1)^n \sum_{l=0}^{n+1} \sum_{k=0}^{n+1-l} (-1)^k C_{n+1}^{k+l} \left(k + \frac{1}{2} - \frac{x}{2}\right)^n \tag{4}$$

has the following properties:

1. $Q'_n(x) = nQ_{n-1}(x)$.
2. $Q_n(x)$ *is even when n is even and odd when n is odd.*
3. $|Q_n(x)| = (-1)^{[n/2]} Q_n(x),\ 0 \leqq x \leqq 1$.
4. $$Q_n(x) = \frac{2^{2n+1}}{n+1}\left[(-1)^{n+1} B_{n+1}\left(\frac{1-x}{4}\right) - B_{n+1}\left(\frac{1+x}{4}\right)\right], \tag{5}$$

where $B_n(x)$ are the Bernoulli polynomials ([2], 1947 *ed., p.* 113).

We introduce the following notation.

Y is the sequence $\{y_m\}$ $(m = 0, \pm 1, \pm 2, \cdots)$.

Δ_p^n is the class of sequences Y satisfying (1).

$f(x, Y)$ is the class of functions defined on the real line and satisfying $f(m) = y_m$ $(m = 0, \pm 1, \pm 2, \cdots)$.

In the present paper we prove the following result.

FUNDAMENTAL THEOREM. *For $1 \leqq p < \infty$ we have*

$$\sup_{Y \in \Delta_p^n} \inf_{f(x) \in f(x, Y)} \| f^{(n)}(x) \|_{L_p} = A_{n,p} \quad (1 \leqslant p < \infty), \tag{6}$$

where

$$A_{n,p} = \frac{(n-1)!}{\left(\int_0^1 |Q_{n-1}(x)|^q \, dx\right)^{1/q}} \quad \left(1 < p < \infty,\ \frac{1}{p} + \frac{1}{q} = 1\right) \tag{7}$$

and

$$A_{n,1} = \frac{(n-1)!}{\max_{0 \leqslant x \leqslant 1} |Q_{n-1}(x)|}. \tag{8}$$

The polynomial $Q_n(x)$ in (7) *and* (8) *is defined by* (5).

We remark that the transformation which corresponds to a sequence of functions satisfying (6) is linear and has the form

$$f^{(n)}(m+t) = Z_{m+1} | Q^*_{n-1}(t) |^{q-1}, \quad 0 < t \leqslant 1$$

for odd n, and

$$f^{(n)}(m+t)=\begin{cases} Z_m\,|\,Q^*_{n-1}(1+t)\,|^{q-1}, & -\frac{1}{2}<t\leqslant 0,\\ Z_m\,|\,Q^*_{n-1}(t)\,|^{q-1}, & 0<t\leqslant\frac{1}{2}\end{cases}$$

for even n, where $Q^*_n(t)$ is defined by (1.4) and

$$Z_m=\sum_{k=-\infty}^{\infty}\sum_{p=1}^{\left[\frac{n}{2}\right]}\frac{x_p^{\left[\frac{n}{2}\right]-1+|k|}}{P'_{2\left[\frac{n}{2}\right]}(x_p)}(n-1)!\,\Delta^n y_{m-\left[\frac{n}{2}\right]+k},$$

and $P_{2[n/2]}(x)$ is defined by (2.4) for odd n and by (2.5) for even n.

We note that when $p\geqq 1$ is fixed and $n\to\infty$, we have the following asymptotic formula for $A_{n,p}$:

$$A_{n,p}=\frac{\pi^{n+\frac{1}{q}}}{2^{n+2}}\left[\frac{\Gamma(q+1)}{\Gamma^2\left(\frac{q+1}{2}\right)}\right]^{1/q}[1+o(1)]. \tag{9}$$

In particular, for $p=1,2,\infty$ we obtain

$$A_{n,1}=\frac{1}{2}\left(\frac{\pi}{2}\right)^n[1+o(1)], \tag{10}$$

$$A_{n,2}=\frac{1}{\sqrt{2}}\left(\frac{\pi}{2}\right)^n[1+o(1)], \tag{11}$$

$$A_{n,\infty}=\frac{\pi}{4}\left(\frac{\pi}{2}\right)^n[1+o(1)]. \tag{12}$$

When $1\leqq p_1\leqq p_2\leqq\infty$, we have the inequalities

$$A_{n,1}\leqslant A_{n,p_1}\leqslant A_{n,p_2}\leqslant A_{n,\infty}.$$

This follows from (7), (8) and the fact that $(\int_0^1|Q_{n-1}(x)|^q dx)^{1/q}$ increases with q.

In §1 we estimate $A_{n,p}$ $(1<p<\infty)$ and obtain the asymptotic behavior of $A_{n,p}$ for fixed p as $n\to\infty$. In §2 we prove lemmas on polynomials, and in §3 we consider the case $p=1$. In §4 we consider a similar problem in several variables.

Let n_1 and n_2 be fixed positive integers, $n=n_1+n_2$; let the values $y_{k,m}$ $(k,m=0,\pm 1,\pm 2,\cdots)$ be assigned at the points with integral coordinates; and let the nth differences

$$\Delta^{n_1+n_2}y_{k,m}=\sum_{l=0}^{n_1}(-1)^{n_1-l}C^l_{n_1}\sum_{s=0}^{n_2}(-1)^{n_2-s}C^s_{n_2}y_{k+l,\,m+s}$$

satisfy

$$\left(\sum_{k=-\infty}^{\infty}\sum_{m=-\infty}^{\infty}|\Delta^{n_1+n_2}y_{k,m}|^p\right)^{1/p}\leqslant 1\quad(1<p<\infty);$$

for the class of sequences $\{y_{k,m}\}$ satisfying this condition we seek the smallest constant $A_{n1,n2,p}$ in the inequality

$$\inf\left(\int_{-\infty}^{\infty} dy \int_{-\infty}^{\infty}\left|\frac{\partial^n f(x,y)}{\partial x^{n_1}\partial y^{n_2}}\right|^p dx\right)^{1/p} \leqslant A_{n_1,n_2,p},$$

where inf is taken over all functions for which $f(k,m) = y_{k,m}$ $(k, m = 0, \pm 1, \cdots)$. We also consider the analogous problem, with the appropriate changes in the hypotheses, for $p=\infty$. We note that the constant $A_{n1,n2,p}$ is connected with $A_{n1,p}$ and $A_{n2,p}$ by

$$A_{n_1,n_2,p} = A_{n_1,p}A_{n_2,p}.$$

Note also that Theorem 1 is known except for the values of the constants [3, 4]. An upper bound (not the smallest possible) for

$$\sup_{Y\in\Delta_p^n}\ \inf_{f(x)\in f(Y)} \|f^{(n)}(x)\|_{L_p}$$

was obtained by a different method by V. S. Rjaben'kiĭ $(p=\infty)$ [5, 6]. S. L. Sobolev [7] showed that Rjaben'kiĭ's interpolation process also furnishes an upper bound for L_p (here also the constant is not exact). Similar remarks hold for functions of several variables.

§1. Bounds for $A_{n,p}$ $(1 < p < \infty)$

We suppose in this section that $1 < p < \infty$ and find upper and lower bounds for

$$f(x) = \sum_{k=0}^{n-1}\frac{f^{(k)}(0)}{k!}x^k + \frac{1}{(n-1)!}\int_0^x (x-t)^{n-1} f^{(n)}(t)\,dt.$$

Let $f(x)\in f(x,Y)$ and let it have an absolutely continuous derivative of order $n-1$. By Taylor's formula with remainder in Cauchy's integral form we have

$$A_{n,p} = \sup_{Y\in\Delta_p^n}\ \inf_{f(x)\in f(x,Y)} \|f^{(n)}(x)\|_{L_p}. \tag{1.1}$$

Proceeding as in [1], we obtain

$$\Delta^n y_m = \frac{1}{(n-1)!}\int_0^1 \sum_{l=0}^{n} f^{(n)}(t+m+l-1)\sum_{k=l}^{n}(-1)^{n-k}C_n^k(k+1-l-t)^{n-1}dt. \tag{1.2}$$

We first find a lower bound for $A_{n,p}$. Let $N > n$ and $f(x)\in f(x,Y)$. Using (1.2), we have

$$\left|\sum_{m=-N}^{N}(-1)^m\Delta^n y_m\right| = \frac{1}{(n-1)!}\left|\sum_{m=-N}^{N}(-1)^m\int_0^1\sum_{l=0}^{n} f^{(n)}(t+m+l-1)\,a_l(t)\,dt\right|,$$

where

$$a_l(t)=\sum_{k=l}^{n}(-1)^{n-k}C_n^k(k+1-l-t)^{n-1}$$

Hence, summing on l and $p=m+l$, we obtain

$$\left|\sum_{m=-N}^{N}(-1)^m\Delta^n y_m\right|=\frac{1}{(n-1)!}\left|\int_0^1\sum_{p=-N+n}^{N}(-1)^p f^{(n)}(t+p-1)\right.$$

$$\times\sum_{l=0}^{n}(-1)^l a_l(t)\,dt+\int_0^1\sum_{p=N+1}^{N+n}(-1)^p f^{(n)}(t+p-1)\sum_{l=p-N}^{n}(-1)^l a_l(t)\,dt$$

$$\left.+\int_0^1\sum_{p=-N}^{-N+n-1}(-1)^p f^{(n)}(t+p-1)\sum_{l=0}^{N+p}(-1)^l a_l(t)\,dt\right|.$$

Let the sequence $\widetilde{Y}=\{\tilde{y}_m\}$ satisfy

$$\Delta^n\tilde{y}_m=\begin{cases}\dfrac{(-1)^m}{(2N+1)^{1/p}}, & |m|\leqslant N,\\ 0, & |m|>N.\end{cases}$$

It is clear that $\widetilde{Y}\in\Delta_p^n$. Now suppose that $f(x)\in f(x,\widetilde{Y})$ and let $M_n=\|f^{(n)}(x)\|_{L_p}$. We may restrict ourselves to the case when M_n is finite. We have

$$\left|\sum_{m=-N}^{N}(-1)^m\Delta^n\tilde{y}_m\right|=(2N+1)^{1-\frac{1}{p}}=(2N+1)^{1/q}$$

Applying Hölder's inequality twice, we obtain

$$\begin{aligned}(2N+1)^{1/q}&\leqslant\frac{1}{(n-1)!}\sum_{m=-N+n-1}^{N-1}\left(\int_0^1|f^{(n)}(m+t)|^p dt\right)^{1/p}\left(\int_0^1|Q_{n-1}^*(t)|^q dt\right)^{1/q}\\&+O(1)M_n\leqslant M_nO(1)+\frac{1}{(n-1)!}\left(\sum_{m=-N+n-1}^{N-1}\int_0^1|f^{(n)}(m+t)|^p dt\right)^{1/p}\\&\times\left(\sum_{m=-N+n-1}^{N=1}\int_0^1|Q_{n-1}^*(t)|^q dt\right)^{1/q}=M_nO(1)+\frac{(2N+1-n)^{1/q}}{(n-1)!}\\&\times\left(\int_0^1|Q_{n-1}^*(t)|^q dt\right)^{1/q}\left(\sum_{m=-N+n-1}^{N-1}\int_0^1|f^{(n)}(m+t)|^p dt\right)^{1/p},\end{aligned}\tag{1.3}$$

where

$$Q_{n-1}^*(t)=\sum_{l=0}^{n}\sum_{k=0}^{l}(-1)^{n-k}C_n^{k+l}(k+1-t)^{n-1}.\tag{1.4}$$

From (1.3) we obtain

$$M_n \geqslant \frac{(n-1)!\,(2N+1)^{1/q}}{(2N+1-n)^{1/q}\left(\int\limits_0^1 |Q^*_{n-1}(t)|^q\,dt\right)^{1/q}+O(1)}\,.$$

Consequently

$$A_{n,p} \geqslant \frac{(n-1)!\,(2N+1)^{1/q}}{(2N+1-n)^{1/q}\left(\int\limits_0^1 |Q^*_{n-1}(t)|^q\,dt\right)^{1/q}+O(1)}$$

for arbitrary N. Letting $N\to\infty$, replacing t by $(1+x)/2$, and using conclusion 2 of Lemma A, we obtain

$$A_{n,p} \geqslant \frac{(n-1)!}{\left|\int\limits_0^1 |Q_{n-1}(x)|^q\,dx\right|^{1/q}}, \tag{1.5}$$

where $Q_n(x)$ is defined by (4).

We now find an upper bound for $A_{n,p}$. We shall show that for every sequence $\{y_k\}$ satisfying (1) there is a function $f^*(x)$ passing through the points (m, y_m) for which

$$\|f^{*(n)}(x)\|_{L_p} \leqslant \frac{(n-1)!}{\left(\int\limits_0^1 |Q_{n-1}(x)|^q\,dx\right)^{1/q}}\,. \tag{1.6}$$

We make the construction as follows: if $n=2s+1$, we look for $f^{(2s+1)}(x)$ in the form

$$f^{(2s+1)}(m+t) = Z_{m+1}\,|Q^*_{2s}(t)|^{q-1}, \quad 0<t\leqslant 1, \tag{1.7}$$

and if $n=2s$, in the form

$$f^{(2s)}(m+t) = \begin{cases} Z_m\,|Q^*_{2s-1}(1+t)|^{q-1}, & -\frac{1}{2}<t\leqslant 0,\\ Z_m\,|Q^*_{2s-1}(t)|^{q-1}, & 0<t\leqslant \frac{1}{2}, \end{cases} \tag{1.8}$$

where $Q^*_n(t)$ is defined by (1.4).

We look for $f^*(x,y)\in f(x,Y)$ in the form (1.1). From (1.2), (1.7) and (1.8) we obtain the following difference equations for Z_m:

$$(2s)!\,\Delta^{2s+1}y_m = \sum_{l=1}^{2s+1} Z_{m+l}\,a_{s,l} \tag{1.9}$$

and

$$(2s-1)!\,\Delta^{2s}y_m = \sum_{l=0}^{2s} Z_{m+l}\,b_{s,l}, \tag{1.10}$$

where

$$a_{s,l} = \int\limits_0^1 |Q^*_{2s}(t)|^{q-1} \sum_{k=l}^{2s+1} (-1)^{k-1} C^k_{2s+1}(k+1-l-t)^{2s}\,dt$$

and

$$b_{s,l} = \int_{1/2}^{1} |Q^*_{2s}(t)|^{q-1} \sum_{k=l}^{2s} (-1)^k C^k_{2s} (k-l+1-t)^{2s-1} dt +$$

$$+ \int_{0}^{1/2} |Q^*_{2s-1}(t)|^{q-1} \sum_{k=l+1}^{2s} (-1)^k C^k_{2s} (k-l-t)^{2s-1} dt.$$

Here the last sum is zero when $l = 2s$.

Let us assume, without proof for the time being, that the difference equation (1.9) satisfies the hypotheses of Theorem 1, including the hypothesis about negative zeros. Then by this theorem we have

$$Z_m = \sum_{k=-\infty}^{\infty} \sum_{n=1}^{s} \frac{x_n^{s-1+|k|}}{P'_{2s}(x_n)} (2s)!\, \Delta^{2s+1} y_{m-s+k}. \tag{1.11}$$

Multiplying (1.11) by $|Z_m|^{p-1} \operatorname{sign} Z_m$ and summing on m from $-N$ to N, we obtain

$$\sum_{m=-N}^{N} |Z_m|^p = \sum_{m=-N}^{N} \sum_{k=-\infty}^{\infty} \sum_{n=1}^{s} \frac{x_n^{s-1+|k|}}{P'_{2s}(x_n)} (2s)!\, \Delta^{2s+1} y_{m-s-k} |Z_m|^{p-1} \operatorname{sign} Z_m$$

$$\leqslant (2s)! \sum_{m=-N}^{N} \sum_{k=-\infty}^{\infty} \left| \sum_{n=1}^{s} \frac{x_n^{s-1+|k|}}{P'_{2s}(x_n)} \right| |\Delta^{2s+1} y_{m-s+k}| |Z_m|^{p-1}.$$

It follows from Lemma 2 of [1] that

$$\left| \sum_{n=1}^{s} \frac{x_n^{s-1+|k|}}{P'_{2s}(x_n)} \right| = (-1)^k \sum_{n=1}^{s} \frac{x_n^{s-1+|k|}}{P'_{2s}(x_n)}$$

Using this, we have

$$\sum_{m=-N}^{N} |Z_m|^p \leqslant (2s)! \sum_{n=1}^{s} \frac{x_n^{s-1}}{P'_{2s}(x_n)} \sum_{k=-\infty}^{\infty} (-x_n)^{|k|} \sum_{m=-N}^{N} |\Delta^{2s+1} y_{m-s+k}| |Z_m|^{p-1}$$

$$\leqslant (2s)! \sum_{n=1}^{s} \frac{x_n^{s-1}}{P'_{2s}(x_n)} \sum_{k=-\infty}^{\infty} (-x_n)^{|k|} \left(\sum_{m=-N}^{N} |\Delta^{2s+1} y_{m-s+k}|^p \right)^{1/p} \left(\sum_{m=-N}^{N} |Z_m|^{(p-1)q} \right)^{1/q}$$

Since $(p-1)q = p$, the last inequality leads to

$$\left(\sum_{m=-N}^{N} |Z_m|^p \right)^{1/p} \leqslant (2s)! \sum_{n=1}^{s} \frac{x_n^{s-1}}{P'_{2s}(x_n)} \sum_{k=-\infty}^{\infty} (-x_n)^{|k|} \left(\sum_{m=-N}^{N} |\Delta^{2s+1} y_{m-s+k}|^p \right)^{1/p}$$

$$\leqslant (2s)! \sum_{n=1}^{s} \frac{x_n^{s-1}}{P'_{2s}(x_n)} \sum_{k=-\infty}^{\infty} (-x_n)^{|k|} \left(\sum_{m=-\infty}^{\infty} |\Delta^{2s+1} y_m|^p \right)^{1/p}$$

$$\leqslant (2s)! \sum_{n=1}^{s} \frac{x_n^{s-1}(1-x_n)}{(1+x_n) P'_{2s}(x_n)} = \frac{(2s)!}{|P_{2s}(-1)|}.$$

The last equation follows from Lemma 1 of [1]. The resulting inequality holds for all N.

Letting $N \to \infty$, we obtain

$$\left(\sum_{m=-\infty}^{\infty} |Z_m|^p\right)^{1/p} \leqslant \frac{(2s)!}{|P_{2s}(-1)|}.$$

It follows from this and (1.7) that

$$\left(\int_{-\infty}^{\infty} |f^{(2s+1)}(t)|^p\, dt\right)^{1/p} = \left(\sum_{m=-\infty}^{\infty} \int_0^1 |Z_m|^p\, |Q_{2s}(t)|^{p(q-1)}\right)^{1/p}$$

$$= \left(\int_0^1 |Q_{2s}^*(t)|^q\, dt\right)^{1/p} \left(\sum_{m=-\infty}^{\infty} |Z_m|^p\right)^{1/p} \leqslant \frac{(2s)!\left(\int_0^1 |Q_{2s}^*(t)|^q dt\right)^{1/p}}{|P_{2s}(-1)|}. \tag{1.12}$$

Using (1.9), we evaluate the characteristic polynomial for $x = -1$. We have

$$P_{2s}(-1) = \sum_{l=1}^{2s+1} (-1)^{l-1} a_{s,l}$$

$$= \int_0^1 |Q_{2s}^*(t)|^{q-1} \sum_{l=1}^{2s+1} (-1)^{l-1} \sum_{k=l}^{2s+1} (-1)^{k+1} C_{2s+1}^k (k+1-l-t)^{2s} dt$$

$$= \int_0^1 |Q_{2s}^*(t)|^{q-1} \sum_{l=0}^{2s+1} \sum_{k=0}^{2s+1-l} (-1)^{k+1} C_{2s+1}^{k+l} (k+1-t)^{2s}\, dt = -\int_0^1 |Q_{2s}^*(t)|^{q-1} Q_{2s}^*(t)\, dt.$$

The last equation follows from (1.4).

If we replace t by $(1+x)/2$, the polynomial $Q_{2s}^*(t)$ becomes $Q_{2s}(x)$, defined in (4). As t varies from 0 to 1, x varies from -1 to 1. It follows from 2 and 3 of Lemma A that $Q_{2s}^*(t)$ does not change sign for $0 \leqq t \leqq 1$.

Hence we have

$$|P_{2s}(-1)| = \int_0^1 |Q_{2s}^*(t)|^q\, dt.$$

From this and (1.12) we obtain

$$\|f^{(2s+1)}(x)\|_{L_p} \leqslant \frac{(2s)!}{\left(\int_0^1 |Q_{2s}^*(t)|^q\, dt\right)^{1/q}}. \tag{1.13}$$

By (1.1), (1.7) and (1.11) we have defined a function whose differences $\Delta^{2s+1}f(m)$ coincide with those of the given sequence $\{y_m\}$. If we require that $f(i) = y_i$ $(i = 0, 1, \cdots, 2s)$, we obtain a function $f^*(x)$ belonging to the class $f(x, Y)$ and satisfying (1.13).

If we replace t by $(1+x)/2$ in (1.13), we obtain

$$\|f^{*(2s+1)}(x)\|_{L_p} \leqslant \frac{(2s)!}{\left(\int_0^1 |Q_{2s}(x)|^q\,dx\right)^{1/q}} \tag{1.14}$$

From (1.5) and (1.14) we obtain (9) for odd n, where $A_{n,p}$ is defined by (7).

Consider the difference equation (1.10). Assuming that the characteristic polynomial of this equation satisfies the hypotheses of Theorem 1, we have

$$Z_m = \sum_{k=-\infty}^{\infty} \sum_{n=1}^{s} \frac{x_n^{s-1+|k|}}{P_{2s}^{*\prime}(x_n)} \Delta^{2s} y_{m-s+k}(2s-1)!, \tag{1.15}$$

where $P_{2s}^*(x)$ is the characteristic polynomial of (1.10). As in the preceding case, we obtain a function $f(x)$ which has the same differences of order $2s$ as the given sequence, and whose nth derivative satisfies

$$\left(\int_{-\infty}^{\infty} |f^{(2s)}(t)|^p\,dt\right)^{1/p} \leqslant \frac{(2s-1)!\left(\int_0^1 |Q_{2s-1}^*(t)|^q\,dt\right)^{1/p}}{|P_{2s}^*(-1)|}. \tag{1.16}$$

If we require that $f(i) = y_i$ $(i = 0, \cdots, 2s-1)$, we obtain a function $f^*(x) \in f(x, Y)$ satisfying (1.16).

Let us calculate the characteristic polynomial of (1.10) at the point -1. We have

$$\begin{aligned}
P_{2s}^*(-1) &= \sum_{l=0}^{2s} (-1)^l \int_{\frac{1}{2}}^{1} |Q_{2s-1}^*(t)|^{q-1} \sum_{k=l}^{2s} (-1)^k C_{2s}^k (k-l+1-t)^{2s-1}\,dt \\
&+ \sum_{l=0}^{2s-1} (-1)^l \int_0^{\frac{1}{2}} |Q_{2s-1}^*(t)|^{q-1} \sum_{k=l+1}^{2s} (-1)^k C_{2s}^k (k-l-t)^{2s-1}\,dt \\
&= \int_{\frac{1}{2}}^{1} |Q_{2s-1}^*(t)|^{q-1} \sum_{l=0}^{2s} \sum_{k=0}^{2s-l} (-1)^k C_{2s}^{k+l} (k+1-t)^{2s-1}\,dt \\
&+ \int_0^{\frac{1}{2}} |Q_{2s-1}^*(t)|^{q-1} \sum_{l=0}^{2s-1} (-1)^l \sum_{k=0}^{2s-l-1} (-1)^{k+l+1} C_{2s}^{k+l+1} (k+1-t)^{2s-1}\,dt \\
&= \int_{\frac{1}{2}}^{1} |Q_{2s-1}^*(t)|^{q-1} Q_{2s-1}^*(t)\,dt - \int_0^{\frac{1}{2}} |Q_{2s-1}^*(t)|^{q-1} Q_{2s-1}^*(t)\,dt.
\end{aligned}$$

In the last equation and in (1.16) we replace t by $(1+x)/2$ and use conclusions 2 and 3 of Lemma A; we obtain

$$\| f^{*(2s)}(x) \|_{L_p} \leqslant \frac{(n-1)!}{\left(\int_0^1 | Q_{2s-1}(x) |^q \, dx \right)^{1/q}}$$

for $f^*(x) \in f(x, Y)$, where $Q_n(x)$ is defined in (4).

From the last inequality and (1.5) we obtain (9) for even n, where $A_{n,p}$ is defined by (7).

We now investigate the asymptotic behavior of $A_{n,p}$ for fixed p as $n \to \infty$.

The Bernoulli polynomials have the following expansions in Fourier series ([2], 1947 ed., p. 113):

$$B_n\left(\frac{x}{2}\right) = \frac{2n!}{(2\pi)^n} \sum_{m=1}^{\infty} \frac{\cos\left(m\pi x - \frac{n\pi}{2}\right)}{m^n}, \quad 0 \leqslant x \leqslant 2. \tag{1.17}$$

Using (5) and (1.17), we obtain

$$| Q_n(x) | = \frac{n!\, 2^{n+2}}{\pi^{n+2}} \left| \sum_{m=0}^{\infty} (-1)^m \frac{\cos\left(\frac{2m+1}{2}\pi x - \frac{n\pi}{2}\right)}{(2m+1)^{n+1}} \right|. \tag{1.18}$$

Using Minkowski's inequality, we obtain from (1.18)

$$\frac{n!\, 2^{n+2}}{\pi^{n+1}} \left[\left(\int_0^1 \left| \cos\left(\frac{\pi x}{2} - \frac{n\pi}{2}\right) \right|^q dx \right)^{\frac{1}{q}} - \left(\int_0^1 \left| \sum_{m=1}^{\infty} (-1)^m \frac{\cos\left(\frac{2m+1}{2}\pi x - \frac{n\pi}{2}\right)}{(2m+1)^{n+1}} \right|^q dx \right)^{\frac{1}{q}} \right]$$

$$\leqslant \left(\int_0^1 | Q_n(x) |^q dx \right)^{1/q} \leqslant \frac{n!\, 2^{n+2}}{\pi^{n+1}} \left[\left(\int_0^1 \left| \cos\left(\frac{\pi x}{2} - \frac{n\pi}{2}\right) \right|^q dx \right)^{1/q} + \left(\int_0^1 \left| \sum_{m=1}^{\infty} (-1)^m \frac{\cos\left(\frac{2m+1}{2}\pi x - \frac{n\pi}{2}\right)}{(2m+1)^{n+1}} \right|^q dx \right)^{1/q} \right].$$

Hence

$$\left(\int_0^1 | Q_n(x) |^q dx \right)^{1/q} = 2n! \left(\frac{2}{\pi}\right)^{n+1} \left(\int_0^1 \sin^q \frac{\pi x}{2} \, dx \right)^{1/q} [1 + o(1)].$$

From this and (7) we have

$$A_{n,p} = \frac{\pi^{n+\frac{1}{q}}}{2^{n+2}} \left[\frac{\Gamma(q+1)}{\Gamma^2\left(\frac{q+1}{2}\right)} \right]^{1/q} [1+o(1)],$$

where $\Gamma(x)$ is the gamma function. This completes the proof of (10).

We say that $\{C_m\} \in l^p$ if $\|C_m\|_{l^p} < \infty$, where

$$\|C_m\|_{l^p} = \begin{cases} \left(\sum_{m=-\infty}^{\infty} |C_m|^p \right)^{1/p} & (1 \leqslant p < \infty), \\ \sup_m |C_m| \ (m = 0, \pm 1, \pm 2, \ldots), & (p = \infty). \end{cases}$$

In obtaining the upper bound for $A_{n,p}$ we have actually obtained a generalization of Theorem 1 of the Introduction. In fact, we have proved

THEOREM 2. *If the polynomial*

$$P_{2n}(x) = \sum_{l=0}^{2n} b_l x^l$$

satisfies

$$P_{2n}(x) \equiv x^{2n} P_{2n}\left(\frac{1}{x}\right),$$

if its zeros are distinct and do not lie on the unit circumference, and if $\{C_m\} \in l^p$, *then the difference equation*

$$\sum_{l=0}^{2n} b_l Z_{m+l} = C_m \quad (m = 0, \pm 1, \pm 2, \ldots) \tag{1.19}$$

has a unique solution $\{Z_m^0\} \in l^p$; *this solution has the representation*

$$Z_m^0 = \sum_{k=-\infty}^{\infty} \sum_{p=1}^{n} \frac{x_p^{n-1+|k|}}{P'_{2n}(x_p)} C_{m-n+k},$$

where x_p $(p = 1, 2, \cdots, n)$ *are the zeros of* $P_{2n}(x)$ *in the disk* $|x| < 1$. *If in addition the zeros are all negative, we have*

$$\|Z_m^0\|_{l^p} \leqslant \frac{\|C_m\|_{l^p}}{|P_{2n}(-1)|}.$$

This bound cannot be improved for the class of sequences $\{C_m\}$ *satisfying* $\|C_m\|_{l^p} \leqq 1$.

We note that the particular solution of (1.19) obtained by the method of variation of parameters [8] (only the values $m = 0, 1, 2, \cdots$ are considered in this reference, but it is also easy to obtain a solution for negative m by the method of variation of parameters) does not belong to l^p, as we can see from Theorem 2.

§2. Properties of the characteristic polynomials

Consider the auxiliary polynomial

$$P_n(t, x) = \sum_{l=0}^{n} x^l \sum_{k=0}^{l} (-1)^k C_{n+1}^k (l-k+t)^n. \tag{2.1}$$

We have the following representation for $P_n(t, x)$:

$$P_n(t, x) = (1-x)^{n+1} g_n(t, x), \tag{2.2}$$

where

$$g_n(t, x) = \sum_{l=0}^{\infty} (l+t)^n x^l. \tag{2.3}$$

In fact,

$$(1-x)^{n+1} \sum_{l=0}^{\infty} (l+t)^n x^l = \sum_{l=0}^{n} x^l \sum_{k=0}^{l} (-1)^k C_{n+1}^k (l+t-k)^n$$
$$+ \sum_{l=n+1}^{\infty} x^l \sum_{k=0}^{n+1} (-1)^k C_{n+1}^k (l-k+t)^n = P_n(t, x),$$

since the second term is identically zero.

Lemma 1. *The characteristic polynomials of* (1.9) *and* (1.10), *namely*

$$P_{2p_1}(x) = \sum_{l=0}^{2p_1} x^l \int_0^1 |Q_{2p_1}^*(t)|^{q-1} \sum_{k=l+1}^{2p_1+1} (-1)^k C_{2p_1+1}^k (k-l-t)^{2p_1} dt \tag{2.4}$$

and

$$P_{2p_1}^*(x) = \sum_{l=0}^{2p_1} x^l \int_{\frac{1}{2}}^1 |Q_{2p_1-1}^*(t)|^{q-1} \sum_{k=l}^{2p_1} (-1)^k C_{2p_1}^k (k-l+1-t)^{2p_1-1} dt$$

$$+ \sum_{l=0}^{2p_1-1} x^l \int_0^{\frac{1}{2}} |Q_{2p_1-1}^*(t)|^{q-1} \sum_{k=l+1}^{2p_1} (-1)^k C_{2p_1}^k (k-l-t)^{2p_1-1} dt \tag{2.5}$$

satisfy the hypotheses of Theorem 1, *i.e. they are symmetric, their zeros are distinct and negative, and* $x = -1$ *is not a zero.*

Proof. We first establish some auxiliary facts. Replacing x by $-e^u$ in (2.3), we obtain

$$g_n(t, x) = e^{-tu} \sum_{l=0}^{\infty} (l+t)^n (-1)^l e^{(l+t)u} = e^{-tu} \frac{d^n}{du^n} \frac{e^{tu}}{1+e^u}. \tag{2.6}$$

Using Rolle's theorem and the fact that

$$\frac{d^n}{du^n} \frac{e^{tu}}{1+e^u} \to 0 \quad \text{as} \quad u \to \pm\infty \quad (0 < t < 1),$$

we can easily show by induction that $g_n(t, x)$, as a function of x, has n distinct negative zeros for $0 < t < 1$.

For $n \geqq 1$ we have

$$g_n(1, x) = e^{-u} \frac{d^n}{du^n} \frac{1 + e^u - 1}{1 + e^u} = -e^{-u} \frac{d^n}{du^n} \frac{1}{1 + e^u} = -e^{-u} g_n(0, x), \quad (2.7)$$

where $x = -e^u$. Therefore $g_n(1, x)$ and $g_n(0, x)$ have the same zeros.

In addition, for $n \geqq 1$ we have

$$g_n(0, x) = \frac{d^n}{du^n} \frac{1}{1 + e^u} = -\frac{d^{n-1}}{du^{n-1}} \frac{e^u}{(1 + e^4)^2} \quad (2.8)$$

and, as before, we find that $g_n(0, x)$ and $g_n(1, x)$ have $n - 1$ distinct negative zeros.

It follows from (2.2) and (2.3) that $P_n(t, x)$ has n distinct negative zeros for $0 < t < 1$, and $P_n(1, x)$ has $n - 1$ distinct negative zeros.

In addition, $P_n(1, x)$ is a polynomial of degree $n - 1$, since it follows from (2.1) that the coefficient of x^n is zero. In fact,

$$\sum_{k=0}^{n} (-1)^k C_{n+1}^k (n + 1 - k)^n = (n + 1) \sum_{k=0}^{n} (-1)^k C_n^k (n + 1 - k)^{n-1} = 0,$$

as the difference of order n of a polynomial of lower degree.

Consider the polynomial (2.4). From the identity

$$\sum_{k=0}^{2p_1+1} (-1)^k C_{2p_1+1}^k (k - l - t)^{2p_1} \equiv 0$$

we have

$$P_{2p_1} = \int_0^1 | Q_{2p_1}^*(t) |^{q-1} \left[\sum_{l=0}^{2p_1} x^l \sum_{k=0} (-1)^k C_{2p_1+1}^k (k - l - t)^{2p_1} \right] dt.$$

From (2.1), (2.2), (2.4), (2.5) and the last equation we obtain

$$P_{2p_1}(-e^u) = \int_0^1 | Q_{2p_1}^*(t) |^{q-1} (1 + e^u)^{2p_1+1} e^{-tu} \frac{d^{2p_1}}{du^{2p_1}} \frac{e^{tu}}{1 + e^u} dt$$

$$= \int_0^{\frac{1}{2}} | Q_{2p_1}^*(t) |^{q-1} (1 + e^u)^{2p_1+1} e^{-tu} \frac{d^{2p_1}}{du^{2p_1}} \frac{e^{tu}}{1 + e^u} dt$$

$$+ \int_{\frac{1}{2}}^1 | Q_{2p_1}^*(t) |^{q-1} (e^u + 1)^{2p_1+1} e^{-tu} \frac{d^{2p_1}}{du^{2p_1}} \frac{e^{tu}}{1 + e^u} dt.$$

In the first integral we replace t by $(1-t)/2$; in the second we replace t by $(1+t)/2$ and use the fact that

$$\left|Q^*_{2p_1}\left(\frac{1+t}{2}\right)\right| = \left|Q^*_{2p_1}\left(\frac{1-t}{2}\right)\right| = |Q_{2p_1}(t)|$$

(the last equation follows from (4) and conclusion 2 of Lemma A). We obtain

$$P_{2p_1}(-e^u) = (1+e^u)^{2p_1+1}\frac{e^{-\frac{u}{2}}}{2}$$

$$\int_0^1 |Q_{2p_1}(t)|^{q-1}\left[e^{tu}\frac{d^{2p_1}}{du^{2p_1}}\frac{e^{\frac{u(1-t)}{2}}}{1+e^u} + e^{-tu}\frac{d^{2p_1}}{du^{2p_1}}\frac{e^{\frac{u(1+t)}{2}}}{1+e^u}\right]dt. \tag{2.9}$$

Since the factor outside the integral does not vanish, and the integrand is an even function of u, it follows that if u_0 is a zero of $P_{2p_1}(-e^u)$ so is $-u_0$. Since $x=-e^u$, the characteristic polynomial (2.4) has a zero at $1/x_0$ if it has one at x_0, i.e. $P_{2p_1}(x)$ is symmetric.

It remains to show that the zeros of (2.4) are negative and distinct, and that $x=-1$ is not a zero. We have already shown that the function $(d^n/du^n)(e^{tu}/(1+e^u))$ has n distinct finite zeros for $0<t<1$, and it follows from (2.1), (2.2), (2.3) and (2.6) that this function has only n zeros, all distinct. Here and later we are considering the zeros of $f(t,u)$ as a function of u for fixed t. In addition, the local maxima of $(d^n/du^n)(e^{tu}/(1+e^u))$ (as a function of u for fixed t) are positive, and the minima are negative.

We now show that as t increases the zeros of $f(t,u)=(d^{2p_1}/du^{2p_1})(e^{tu}/(1+e^u))$ also increase. The equation $f(t,u)=0$ defines u as a function of t. Consider an arbitrary t_0, $0<t_0<1$. From what was proved above, there are only $2p_1$ points u_i^0 $(i=1,\cdots,2p_1)$ at which $f(t_0,u_i^0)=0$. The function $f(t,u)$ is continuous together with its partial derivatives $f'_t(t,u)$ and $f'_u(t,u)$, and $f'_u(t_0,u_i^0)\neq 0$, since the zeros are simple. Hence by the implicit function theorem (see for example [**9**]) there are, in neighborhoods of (t_0,u_i^0), single-valued functions $u_i=u_i(t)$ such that $u_i^0=u_i(t_0)$ and $u_i=u_i(t)$ have continuous derivatives in neighborhoods of (t_0,u_i^0), and these derivatives are given by the formula

$$u'_i(t) = -\frac{f'_t(t,u)}{f'_u(t,u)} = -\frac{u\dfrac{d^{2p_1}}{du^{2p_1}}\dfrac{e^{tu}}{1+e^u} + 2p_1\dfrac{d^{2p_1-1}}{du^{2p_1-1}}\dfrac{e^{tu}}{1+e^u}}{\dfrac{d^{2p_1+1}}{du^{2p_1+1}}\dfrac{e^{tu}}{1+e^u}} \tag{2.10}$$

Putting $t=t_0$ and $u=u_i^0$ in (2.10), and using the facts that the zeros of $(d^{2p_1}/du^{2p_1})(e^{t_0u}/(1+e^u))$ are extrema of $(d^{2p_1-1}/du^{2p_1-1})(e^{t_0u}/(1+e^u))$ and that the local minima of the latter function are negative, while the maxima are positive, we obtain $u'_i(t_0)>0$ for $0<t_0<1$, i.e. the zeros of $f(t,u)$ increase with t.

Consider the function defined by (2.6),

$$g_{2p_1}\left(\frac{1+t}{2}, x\right) = e^{-\frac{u(1+t)}{2}} \frac{d^{2p_1}}{du^{2p_1}} \frac{e^{\frac{u}{2}}}{1+e^u} e^{\frac{tu}{2}}, \tag{2.11}$$

where $x = e^{-u}$ and $0 \leqq t < 1$.

We have already shown that $g_{2p_1}((1+t)/2, x)$ has $2p_1$ distinct zeros for $0 \leqq t < 1$. Let $-\infty < u_1(t) < \cdots < u_{2p_1}(t) < \infty$ be these zeros. We have shown that $u_i(t)$ $(i = 1, \cdots, 2p_1)$ increase with t. We have also shown that $g_{2p_1}(1, x)$ has $2p_1 - 1$ zeros, all distinct. Let $-\infty < u_1(1) < \cdots < u_{2p_1-1}(1) < \infty$ be the zeros of $g_{2p_1}(1, x)$. Since $g_{2p_1}(t, u)$ is continuous together with its partial derivatives $\partial g_{2p_1}(t, u)/\partial t$ and $\partial g_{2p_1}(t, u)/\partial u$ in a neighborhood of the point $[1, u_i(1)]$ $(i = 1, \cdots, 2p_1 - 1)$ and

$$\frac{\partial}{\partial u} g_{2p_1}[1, u_i(1)] \neq 0 \qquad (i = 1, 2, \ldots, 2p_1 - 1),$$

since the zeros are simple, it follows that in a neighborhood $1 - \epsilon < t < 1 + \epsilon$ there are single-valued functions $u_k = u_k^0(t)$ such that $g_{2p_1}[t, u_k^0(t)] = 0$ and $u_k^0(1) = u_k(1)$. But in a neighborhood $0 < t < 1$ the functions $u_k^0(t)$ $(k = 1, \cdots, 2p - 1)$ coincide with the $2p - 1$ functions $u_i(t)$ $(i = 1, \cdots, 2p)$ already defined; moreover, $u_k^0(t) = u_k(t)$ $(k = 1, \cdots, 2p - 1)$, and $u_{2p}(t)$ tends to ∞ as $t \to 1 - 0$, since $u_{2p}(t)$ increases with t, and if $u_{2p}(t)$ had a finite limit as $t \to 1 - 0$, then $u_i(t)$ $(i < 2p)$ would also have finite limits, since they increase with t and would be bounded above. From this and the continuity of $g_{2p}(t, x)$ it would follow that $g_{2p}(1, x)$ has $2p$ zeros, which is impossible. Consequently we have

$$u_i(0) \leqslant u_i(t) \leqslant u_i(1) \quad (i = 1, \ldots, 2p - 1), \ 0 \leqslant t \leqslant 1 \tag{2.12}$$

Since the zeros of $g_{2p_1}(t, x)$ increase with t, and the zeros of $g_{2p_1}(0, x)$ and $g_{2p_1}(1, x)$ are the same, it follows that the zeros of $g_{2p_1}(\frac{1}{2}, x)$ and $g_{2p_1}(1, x)$ alternate, i.e.

$$u_1(0) < u_1(1) < u_2(0) < \ldots < u_{2p-1}(1) < u_{2p}(0). \tag{2.13}$$

From (2.12) and (2.13) we obtain

$$u_i(t) \leqslant u_i(1) < u_{i+1}(0) \leqslant u_{i+1}(t) \qquad (i = 1, \ldots, 2p - 1). \tag{2.14}$$

We have a similar result for

$$g_{2p_1}\left(\frac{1-t}{2}, x\right) = e^{\frac{u(t-1)}{2}} \frac{d^{2p_1}}{du^{2p_1}} \left[\frac{e^{\frac{u}{2}}}{1+e^u} e^{-\frac{tu}{2}}\right], \tag{2.15}$$

where $x = -e^u$ and $0 \leqq t < 1$. Let $-\infty < \bar{u}_2(t) < \cdots < \bar{u}_{2p_1}(t) < \infty$ be the zeros of $g_{2p_1}(1 - t)/2, x)$.

It follows from what we have already proved that $\bar{u}_i(t)$ $(i = 1, \cdots, 2p_1)$

decrease as t increases. We have also shown that $g_{2p_1}(0,x)$ has only $2p_1-1$ finite zeros and that the zeros of $g_{2p_1}(0,x)$ and $g_{2p_1}(1,x)$ are the same. Hence, as before, we obtain

$$u_i(1) \leqslant \bar{u}_{i+1}(t) \leqslant \bar{u}_{i+1}(0) \quad (i=1,\ldots,2_{p_1}-1), \ 0 \leqslant t < 1. \tag{2.16}$$

In addition, the functions defined by (2.11) and (2.15) for $t=0$ are the same. Hence we obtain

$$u_i(0) = \bar{u}_i(0) \quad (i=1,\ldots, 2p_1). \tag{2.17}$$

It follows from (2.2) that

$$P_{2p_1}(t, 0) > 0, \quad 0 < t \leqslant 1. \tag{2.18}$$

Since $x = -e^u$, the point $x=0$ corresponds to $u=-\infty$. It follows from (2.2), (2.3), (2.6) and (2.18) that the function defined by (2.11) is positive for all t, $0 \leqq t \leqq 1$, in a neighborhood of $u=-\infty$. The same is true for $0 \leqq t \leqq 1$ for the function defined by (2.15). Since $g_{2p_1}((1+t)/2, x)$ and $g_{2p_1}((1-t)/2, x)$ have only simple zeros, we have

$$\operatorname{sign} g_{2p_1}\left(\frac{1+t}{2}, x\right) = (-1)^i \quad (i=1,\ldots,2p_1-1) \tag{2.19}$$

for $u_i(t) < u < u_{i+1}(t)$, $0 \leqq t < 1$, and

$$\operatorname{sign} g_{2p_1}\left(\frac{1-t}{2}, x\right) = (-1)^i \ (i=1,\ldots,2p_1-1)\ (x=-e^u) \tag{2.20}$$

for $\bar{u}_i(t) < u < \bar{u}_{i+1}(t)$, $0 \leqq t < 1$.

Now let $x_i = -e^{u_i(1)}$ $(i=1,\cdots,2p_1-1)$. From (2.19) and (2.14) we obtain

$$\operatorname{sign} g_{2p_1}\left(\frac{1+t}{2}, x_i\right) = (-1)^i, \ 0 < t < 1. \tag{2.21}$$

From (2.17), (2.13) and (2.16) we obtain

$$\bar{u}_i(t) < u_i(1) < \bar{u}_{i+1}(t), \quad 0 \leqslant t < 1.$$

From this and (2.20) we obtain

$$\operatorname{sign} g_{2p_1}\left(\frac{1-t}{2}, x_i\right) = (-1)^i \ (i=1,\ldots,2p_1-1), \ 0 < t < 1. \tag{2.22}$$

Therefore from (2.22), (2.21), (2.15), (2.11) and (2.9) we obtain

$$\operatorname{sign} P_{2p_1}(x_i) = (-1)^i \ (i=1,\ldots, 2p_1-1).$$

It follows from (2.4) that $P_{2p_1}(0) > 0$, and since the coefficient of x^{2p_1} is equal to the constant term, we have $P_{2p_1}(x) \to +\infty$ as $x \to -\infty$; consequently there is a point $x_{2p_1} < \min_{1 \leqq i \leqq 2p_1-1} x_i$ at which $P_{2p_1}(x_{2p_1}) > 0$. Thus we have $2p_1+1$ points at which $P_{2p_1}(x)$ takes alternating signs, and consequently it has $2p_1$ distinct zeros, all negative. In addition, $x = -1$ is not a zero of $P_{2p_1}(x)$, since it follows from (2.7) and (2.8) that $e^u g_{2p_1}(1, -e^u)$ is an odd function of u and hence $u = 0$ coincides with one of the values $u_i(1)$. But $\operatorname{sign} P_{2p_1}[-e^{u_i(1)}] = (-1)^i \neq 0$.

We have therefore proved that the characteristic polynomial $P_{2p_1}(x)$ of (1.9) satisfies all the hypotheses of Theorem 1, and the assumption made just before (1.11) is justified.

Now consider the characteristic polynomial of (1.10). Using the identity

$$\sum_{k=0}^{2p_1} (-1)^k C_{2p_1}^k (k - l + t)^{2p_1-1} \equiv 0$$

and (2.5), we obtain

$$\begin{aligned} P^*_{2p_1}(x) = {} & \sum_{l=1}^{2p_1} x^l \int_{1/2}^{1} |Q^*_{2p_1-1}(t)|^{q-1} \sum_{k=0}^{l-1} (-1)^k C_{2p_1}^k (l - k - 1 + t)^{2p_1-1}\, dt \\ & + \sum_{l=0}^{2p_1-1} x^l \int_0^{1/2} |Q^*_{2p_1-1}(t)|^{q-1} \sum_{k=0}^{l} (-1)^k C_{2p_1}^k (l - k + t)^{2p_1-1}\, dt. \end{aligned}$$

In the first integral replace t by $(1 + t)/2$; in the second, replace t by $(1 - t)/2$; we obtain

$$\begin{aligned} P^*_{2p_1}(x) = {} & \frac{1}{2} \sum_{l=1}^{2p_1} x^l \int_0^1 \left| Q^*_{2p_1-1}\left(\frac{1+t}{2}\right)\right|^{q-1} \sum_{k=0}^{l-1} (-1)^k C_{2p_1}^k \left(l - k - \frac{1-t}{2}\right)^{2p_1-1} dt \\ & + \frac{1}{2} \sum_{l=0}^{2p_1-1} x^l \int_0^1 \left| Q^*_{2p_1-1}\left(\frac{1-t}{2}\right)\right|^{q-1} \sum_{k=0}^{l} (-1)^k C_{2p_1}^k \left(l - k + \frac{1-t}{2}\right)^{2p_1-1} dt. \end{aligned}$$

Since $Q^*_{2p-1}((1 + t)/2)$ is equal to the polynomial $Q_{2p-1}(t)$ defined by (4), by using conclusion 2 of Lemma A we finally obtain

$$\begin{aligned} P_{2p_1}(x) = {} & \frac{1}{2} \int_0^1 |Q_{2p_1-1}|^{q-1} \Bigg[\sum_{l=0}^{2p_1-1} x^l C_{2p_1}^k \left(l - k + \frac{1-t}{2}\right)^{2p_1-1} \\ & + x \sum_{l=1}^{2p_1} x^{l-1} \sum_{k=0}^{l-1} (-1)^k C_{2p_1}^k \left(l - 1 - k + \frac{1-t}{2}\right)^{2p_1-1} \Bigg] dt \\ = {} & \frac{1}{2} \int_0^1 |Q_{2p_1-1}(t)|^{q-1} \Bigg[\sum_{l=0}^{2p_1-1} x^l \sum_{k=0}^{l} (-1)^k C_{2p_1}^k \left(l - k + \frac{1-t}{2}\right)^{2p_1-1} \\ & + x \sum_{l=0}^{2p_1-1} x^l \sum_{k=0}^{l} (-1)^k C_{2p_1}^k \left(l - k + \frac{1+t}{2}\right)^{2p_1-1} \Bigg] dt. \end{aligned} \tag{2.23}$$

We transform the right-hand side by using (2.1), (2.2), (2.3) and (2.6). We have

$$P^*_{2p}(x) = \frac{(1+e^u)^{2p}}{2}\int_0^1 |Q_{2p-1}(t)|^{q-1}\left[e^{\frac{t-1}{2}u}\frac{d^{2p-1}}{du^{2p-1}}\frac{e^{\frac{u(1-t)}{2}}}{1+e^u}\right.$$
$$\left.-e^{\frac{1-t}{2}u}\frac{d^{2p-1}}{du^{2p-1}}\frac{e^{\frac{u(1+t)}{2}}}{1+e^u}\right]dt, \quad x=-e^u. \tag{2.24}$$

It is clear from (2.24) that if u_0 is a zero of $P^*_{2p}(-e^u)$ then $-u_0$ is also a zero. Consequently $P^*_{2p}(x)$ is a symmetric polynomial.

As in the preceding case, it is easily established that

$$f_1(t,u) = e^{\frac{t-1}{2}u}\frac{d^{2p-1}}{du^{2p-1}}\frac{e^{\frac{u(1-t)}{2}}}{1+e^u} \tag{2.25}$$

and

$$f_2(t,u) = e^{\frac{1-t}{2}u}\frac{d^{2p-1}}{du^{2p-1}}\frac{e^{\frac{u(1+t)}{2}}}{1+e^u} \tag{2.26}$$

have $2p-1$ zeros for $0 \leqq t < 1$, and that the zeros of $f_1(t,u)$ decrease as t increases, whereas the zeros of $f_2(t,u)$ increase. Since the zeros of $g_{2p_1-1}(t,x)$ increase, and the zeros of $g_{2p_1-1}(0,x)$ and of $g_{2p_1-1}(1,x)$ are the same, it follows that the zeros of $f_k(0,u)$ and $f_k(1,u)$ $(k=1,2)$ alternate.

Let $u_i(t)$ $(i=1,\cdots,2p_1-1)$ be the zeros of $f_1(t,u)$, arranged in increasing order, and let $\bar{u}_i(t)$ $(i=1,\cdots,2p_1-1)$ be the zeros of $f_2(t,u)$, also arranged in increasing order. It is clear that $u_i(0)=\bar{u}_i(0)$ $(i=1,\cdots,2p_1-1)$. It follows from (2.6), (2.7) and (2.8) that $2p_1-2$ of the zeros of $f_1(1,u)$ and $f_2(1,u)$ coincide, i.e. $u_{i+1}(1)=\bar{u}_i(1)$ $(i=1,\cdots,2p_1-2)$, whereas $u_1(1)=-\infty$, $\bar{u}_{2p_1-1}(1)=\infty$.

It follows from these properties that

$$u_i(t) \leqslant u_i(0) < u_{i+1}(1) \leqslant u_{i+1}(t) \leqslant u_{i+1}(0) \quad (i=1,\ldots,2p_1-2) \tag{2.27}$$

and

$$u_i(0) \leqslant \bar{u}_i(t) \leqslant \bar{u}_i(1) < u_{i+1}(0) \leqslant u_{i+1}(t) \quad (i=1,\ldots,2p_1-2). \tag{2.28}$$

It is easily established by using (2.23) and (2.24) that, for $0 \leqq t < 1$, the functions $f_1(t,u)$ and $f_2(t,u)$ are positive in a neighborhood of $u=-\infty$ ($x=0$ corresponds to $u=-\infty$). Since the zeros of $f_1(t,u)$ and $f_2(t,u)$ are simple, we have

$$\operatorname{sign} f_1(t,u) = (-1)^i, \quad u_i(t) < u < u_{i+1}(t) \quad (i=1,\ldots,2p_1-2) \tag{2.29}$$

for $0<t<1$, and

$$\operatorname{sign}[-f_2(t,u)] = (-1)^{i+1}, \quad \bar{u}_i(t) < u < \bar{u}_{i+1}(t) \quad (i=1,\ldots,2p_1-2). \tag{2.30}$$

It follows from (2.29), (2.30), (2.27) and (2.28) that

$$\operatorname{sign}\left[e^{\frac{t-1}{2}u}\frac{d^{2p_1-1}}{du^{2p_1-1}}\frac{e^{\frac{u(1-t)}{2}}}{1+e^u}-e^{\frac{1-t}{2}u}\frac{d^{2p_1-1}}{du^{2p_1-1}}\frac{e^{\frac{u(1+t)}{2}}}{1+e^u}\right]_{u=u_i(0)}=(-1)^i. \quad (2.31)$$

$$(i=2,\dots,\ 2p_1-2).$$

Consider the points $u=u_1(0)$ and $u=u_{2p_1-1}(0)$. For $u=u_1(0)$ it follows from (2.27) and (2.29) that $\operatorname{sign} f_1[t,u_1(0)]=-1$, $0<t<1$, and from (2.28) and (2.30) we obtain $\operatorname{sign} f_2[t,u_1(0)]=1$. Hence it follows from (2.23) and (2.26) that (2.31) holds for $i=1$.

In the case when $u=u_{2p_1-1}(0)$ it follows from (2.27) and (2.29) that $\operatorname{sign} f_1[t,u_{2p_1-1}(0)]=1$, and from (2.28) and (2.30) we obtain $\operatorname{sign} f_2[t,u_{2p_1-1}(0)]=1$. From this and (2.25), (2.26) it follows that (2.31) holds for $i=2p_1-1$. Thus

$$\operatorname{sign}\left[e^{\frac{t-1}{2}u}\frac{d^{2p_1-1}}{du^{2p_1-1}}\frac{e^{\frac{u(1-t)}{2}}}{1+e^u}-e^{\frac{1-t}{2}u}\frac{d^{2p_1-1}}{du^{2p_1-1}}\frac{e^{\frac{u(1+t)}{2}}}{1+e^u}\right]_{u=u_i(0)}=(-1)^i \quad (2.32)$$

$$(i=1,\dots,2p_1-1).$$

Let $x_i=-e^{u_i(0)}$; then from (2.32) and (2.24) we obtain

$$\operatorname{sign} P^*_{2p}(x_i)=(-1)^i \quad (i=1,\dots,\ 2p_1-1);$$

here $-\infty<x_{2p_1-1}<\cdots<x_1<0$ and one x_i is -1, i.e. $x=-1$ is not a zero of $P^*_{2p_1}(x)$. It is clear from (2.23) that $P^*_{2p_1}(0)>0$.

Since $P^*_{2p_1}(x)$ is a symmetric polynomial, the coefficient of x^{2p_1} is positive; consequently there is a point $x_{2p_1}<x_{2p_1-1}$ at which $P^*_{2p_1}(x_{2p_1})>0$. Thus there are points $-\infty<x_{2p_1}<x_{2p_1-1}<\cdots<x_1<x_0=0$ at which $P_{2p_1}(x)$ takes values with alternating signs. It follows that the zeros of $P^*_{2p_1}(x)$ are all real, negative, and distinct, and that $P^*_{2p_1}(-1)\neq 0$.

This completes the proof of the validity of the assertion used in solving (1.10). In proving Lemma 2 we used only the following property of $Q^*_n(t)$:

$$|Q^*_n(t)|=|Q^*_n(1-t)|,\quad 0\leqslant t\leqslant\frac{1}{2}.$$

Hence, if we repeat the reasoning of Lemma 3, we obtain

Theorem 3. *If the polynomial $P_{2n}(x)$ is representable in the form*

$$\frac{P_{2n}(x)}{(1-x)^{2n+1}}=\int_0^1 f(t)\sum_{l=0}^{\infty}(l+t)^{2n}x^l dt \quad (2.33)$$

or in the form

$$\frac{P_{2n}(x)}{(1-x)^{2n}}=\int_0^{1/2} f(t)\sum_{l=0}^{\infty}(l+t)^{2n-1}x^l dt+x\int_{1/2}^{1} f(t)\sum_{l=0}^{\infty}(l+t)^{2n-1}x^l dt.$$

where $f(t)$ is nonnegative, integrable, different from zero on a set of positive measure, and satisfies $f(t) = f(1-t)$, $0 \leqq t \leqq \frac{1}{2}$, then $P_{2n}(x) \equiv x^{2n} P_{2n}(1/x)$, its zeros are negative and distinct, and $x = -1$ is not a zero of $P_{2n}(x)$.

The special case of (2.33) when $f(t) \equiv 1$ was considered in [**10**].

§3. The case $p = 1$

In the present section we consider the case $p = 1$.

We shall show that there is a sequence $Y \in \Delta_1^n$ such that for each $f(x) \in f(x, Y)$

$$\int_{-\infty}^{\infty} |f^{(n)}(x)|\, dx = \| f^{(n)}(x) \|_{L_1} \geqslant \frac{(n-1)}{\max\limits_{0 \leqslant t \leqslant 1} | Q_{n-1}(t) |}. \tag{3.1}$$

Since $f^{(n)}(x) \in L_1(-\infty, \infty)$ it follows that $\int_k^{k+1} |f^{(n)}(x)|\, dx \to 0$ as $|k| \to \infty$. Let

$$a_l(t) = \sum_{k=l}^{n} (-1)^{n-k} C_n^k (k+1-l-t)^{n-1}.$$

To obtain a lower bound, we use (1.2) and form the sum

$$\begin{aligned} \left| \sum_{m=-N}^{N} (-1)^m \Delta^n y_m \right| &= \frac{1}{(n-1)!} \left| \sum_{m=-N}^{N} (-1)^m \int_0^1 \sum_{l=0}^{n} f^{(n)}(t+m+l-1) a_l(t)\, dt \right| \\ &= \frac{1}{(n-1)!} \left| \sum_{m=-N}^{N} \int_0^1 (-1)^m f^{(n)}(m+t) \sum_{l=0}^{n} (-1)^l a_l(t)\, dt \right| + o(1) \\ &\leqslant \frac{1}{(n-1)!} \max_{0 \leqslant t \leqslant 1} \left| \sum_{l=0}^{n} (-1)^l a_l(t) \right| \sum_{m=-N}^{N} \int_0^1 | f^{(n)}(m+t) |\, dt + o(1). \end{aligned} \tag{3.2}$$

If we put $\Delta^n y_m = 1$ for $m = 0$ and $\Delta^n y_m = 0$ for $m \neq 0$ and take the limit as $N \to \infty$, we obtain (3.1). Hence we have

$$\sup_{Y \in \Delta_1^n} \inf_{f(x) \in f(x, Y)} \| f^{(n)}(x) \|_{L_1} \geqslant \frac{(n-1)!}{\max\limits_{0 \leqslant t \leqslant 1} | Q_{n-1}(t) |}. \tag{3.3}$$

We can show that for each $\epsilon > 0$ and every sequence $Y \in \Delta_1^n$ there is a function $f^*(x) \in f(x, Y)$ such that

$$\| f^{*(n)}(x) \|_{L_1} < \frac{(n-1)!}{\max\limits_{0 \leqslant t \leqslant 1} | Q_{n-1}(t) |} + \varepsilon. \tag{3.4}$$

It is enough to take

$$f_\delta^{(2p)}(m+t) = \begin{cases} \dfrac{Z_m}{2\delta}, & t \in [-\delta, \delta] \\ 0, & t \overline{\in} [-\delta, \delta] \\ \left(-\frac{1}{2} \leqslant t \leqslant \frac{1}{2},\ 0 < \delta < \frac{1}{2},\ m = 0, \pm 1, \ldots \right) \end{cases}$$

$$f_\delta^{(2p+1)}(m+t)=\begin{cases}\dfrac{Z_m}{\delta}, & t\in[0,\delta]\\ 0, & t\overline{\in}[0,\delta]\\ (0\leqslant t\leqslant 1,\ 0<\delta<1,\ m=0,\pm 1,\dots).\end{cases}$$

We determine Z_m from a difference equation whose characteristic polynomial satisfies the hypotheses of Theorem 1. We have the following estimate:

$$\int_{-\infty}^{+\infty}|f_\delta^{*(n)}(x)|\,dx=\sum_{m=-\infty}^{\infty}|Z_m|\leqslant\frac{\delta(n-1)!}{P_n(\delta)},$$

where $P_n(\delta)/\delta\to\max_{0\leq t\leq 1}|Q_{n-1}(t)|$ as $\delta\to 0$. it follows that for each $\epsilon>0$ there is a $\delta(\epsilon)$ such that (3.4) holds for $f_\delta^{*(n)}(x)$. From (3.4) and (3.3) we have

$$\sup_{Y\in\Delta_1^n}\ \inf_{f(x)\in f(x,Y)}\|f^{(n)}(x)\|_{L_1}=\frac{(n-1)!}{\max\limits_{0\leqslant t\leqslant 1}|Q_{n-1}(t)|}.\tag{3.5}$$

As $\delta\to 0$, the limit of $f_\delta^{(n)}(x)$ is a function not belonging to L_1. Hence we shall consider a wider class of functions, namely those for which the $(n-1)$th derivative has bounded variation on the whole line.

We seek the smallest value, for the class of sequences $\{y_n\}$ satisfying (1) for $p=1$, of the constant $A_{n,1}$ in the inequality

$$\inf\bigvee_{-\infty}^{\infty}f^{(n-1)}(x)=\inf\int_{-\infty}^{\infty}|df^{(n-1)}(x)|\leqslant A_{n,1},$$

where inf is taken over all functions for which $f(k)=y_k$ $(k=0,\pm 1,\pm 2,\cdots)$.

By Taylor's formula with remainder in Cauchy's integral form we have

$$f(x)=\sum_{k=0}^{n}\frac{f^{(k)}(x_0)}{k!}(x-x_0)^k+\frac{1}{(n-1)!}\int_{x_0}^{x}(x-t)^{n-1}\,df^{(n-1)}(x).\tag{3.6}$$

As in the L_1 case we find that for every $f(x)\in f(x,\tilde{Y})$ and every sequence $\tilde{Y}$ for which $\Delta^n\tilde{y}_m=1$ for $m=0$ and $\Delta^n\tilde{y}_m=0$ for $m\neq 0$ we have the inequality

$$\bigvee_{-\infty}^{\infty}f^{(n-1)}(x)=\int_{-\infty}^{\infty}|df^{(n-1)}(x)|\geqslant\frac{(n-1)!}{\max\limits_{0\leqslant t\leqslant 1}|Q_{n-1}(t)|},$$

i.e.

$$A_{n,1}=\sup_{Y\in\Delta_1^n}\ \inf_{f(x)\in f(x,y)}\bigvee_{-\infty}^{\infty}f^{(n-1)}(x)\geqslant\frac{(n-1)!}{\max\limits_{0\leqslant t\leqslant 1}|Q_{n-1}(t)|}.\tag{3.7}$$

We now obtain an upper bound for $A_{n,1}$. We shall show that for every sequence $Y\in\Delta_1^n$ there is a function $f^*(x)$ belonging to $f(x,Y)$ for which

$$\bigvee_{-\infty}^{\infty} f^{*\,(n-1)}(x) \leqslant \frac{(n-1)!}{\max\limits_{0\leqslant t\leqslant 1} |Q_{n-1}(t)|}. \tag{3.8}$$

Using (3.6) and arguing as in [1], we obtain

$$\Delta^n y_m = \frac{1}{(n-1)!}$$
$$\times \int_0^1 \sum_{l=0}^{n} \sum_{k=l}^{n} (-1)^{n-k} C_n^k (k+1-l-t)^{n-1} \times df^{(n-1)}(t+m+l-1). \tag{3.9}$$

We make the construction as follows: if $n = 2p$, we put

$$f^{(2p-1)}(m+t) = \alpha_m, \quad 0 \leqslant t < 1, \quad m = 0, \pm 1, \pm 2, \ldots, \tag{3.10}$$

and if $n = 2p+1$ we put

$$f^{(2p)}(m+t) = \alpha_m, \quad -\frac{1}{2} \leqslant t \leqslant \frac{1}{2}, \quad m = 0, \pm 1, \pm 2, \ldots. \tag{3.11}$$

Using (3.9), (3.10) and (3.11), we obtain the difference equations

$$\sum_{l=0}^{2p} Z_{m+l} \sum_{k=0}^{l} (-1)^k C_{2p+1}^k \left(l-k+\frac{1}{2}\right)^{2p} = (2p)!\, \Delta^{2p+1} y_m \tag{3.12}$$

and

$$\sum_{l=1}^{2p} Z_{m+l-1} \sum_{k=l}^{2p} (-1)^k C_{2p}^k (k-l)^{2p-1} = (2p-1)!\, \Delta^{2p} y_m, \tag{3.13}$$

where $Z_m = \alpha_{m+1} - \alpha_m$.

Consider the characteristic polynomial of (3.12). We have

$$P_{2p}(x) = \sum_{l=0}^{2p} x^l \sum_{k=0}^{l} (-1)^k C_{2p+1}^k \left(l-k+\frac{1}{2}\right)^{2p}$$
$$= (1-x)^{2p+1} \sum_{l=0}^{\infty} \left(l+\frac{1}{2}\right)^{2p} x^l,$$

or, putting $x = -e^u$,

$$P_{2p}(-e^u) = (1+e^u)^{2p+1} e^{-\frac{u}{2}} \frac{d^{2p}}{du^{2p}} \frac{e^{\frac{u}{2}}}{1+e^u}.$$

It is then clear that (3.12) satisfies the hypotheses of Theorem 1. Consequently

$$Z_m = \sum_{k=-\infty}^{\infty} \sum_{n=1}^{p} \frac{x_n^{p-1+|k|}}{P'_{2p}(x_n)} (2p)!\, \Delta^{2p+1} y_{m-p+k}$$

and

$$\sum_{m=-N}^{N} |Z_m| = \int_{-N}^{N} |df^{(n-1)}(t)| =$$

$$= \sum_{m=-N}^{N} \left| \sum_{k=-\infty}^{\infty} \sum_{n=1}^{p} \frac{x_n^{p-1+|k|}}{P'_{2p}(x_n)} (2p)!\, \Delta^{2p+1} y_{m-p+k} \right|$$

$$\leqslant \sum_{k=-\infty}^{\infty} (-1)^k \sum_{n=1}^{p} \frac{x_n^{p-1+|k|}}{P'_{2p}(x_n)} \sum_{m=-N}^{N} |\Delta^{2p+1} y_{m-p+k}|\,(2p)! \tag{3.14}$$

$$\leqslant \frac{(2p)!}{P_{2p}(-1)} = \frac{(2p)!}{\left|Q^*\left(\frac{1}{2}\right)\right|} = \frac{(2p)!}{\max\limits_{0\leqslant t\leqslant 1} |Q_{2p}(t)|}.$$

Consider the characteristic polynomial of (3.13). We have

$$P^*_{2p-2}(x) = \sum_{l=0}^{2p-1} x^l \sum_{k=l+1}^{2p} (-1)^k C^k_{2p} (k-1-l)^{2p-1}$$

$$= \sum_{l=0}^{2p-1} x^l \sum_{k=0}^{l} (-1)^k C^k_{2p} (l+1-k)^{2p-1}.$$

We shall show that the coefficient of x^{2p-1} is zero. In fact,

$$\sum_{k=0}^{2p-1} (-1)^k C^k_{2p} (2p-k)^{2p-1} = 2p \sum_{k=0}^{2p-1} (-1)^k C^k_{2p-1} (2p-k)^{2p-2} = 0,$$

as the difference of order $2p-1$ of a polynomial of lower degree. In addition,

$$P^*_{2p-2}(x) = \sum_{l=0}^{2p-2} x^l \sum_{k=0}^{l} (-1)^k C^k_{2p} (l+1-k)^{2p-1}$$

$$= (1-x)^{2p} \sum_{l=0}^{\infty} (l+1)^{2p-1} x^l = (1+e^u)^{2p} e^{-u} \frac{d^{2p-1}}{du^{2p-1}} \frac{e^u}{1+e^u}$$

$$= (1+e^u)^{2p} e^{-u} \frac{d^{2p-2}}{du^{2p-2}} \frac{e^u}{(1+e^u)^2}, \quad x = -e^u.$$

It follows that the characteristic polynomial of (3.13) satisfies the hypotheses of Theorem 1, and consequently the solution of (3.13) has the form

$$Z_m = \sum_{k=-\infty}^{\infty} \sum_{n=1}^{p-1} \frac{x_n^{p-2+|k|}}{P^{*\prime}_{2p-2}(x_n)} (2p-1)!\, \Delta^{2p} y_{m-p+k}$$

and

$$\int_{-N}^{N} |df^{(2p-1)}(t)| = \sum_{m=-N}^{N} |Z_m|$$

$$= \sum_{m=-N}^{N} \left| \sum_{k=-\infty}^{\infty} \sum_{n=1}^{p-1} \frac{x_n^{p-2+|k|}}{P^{*\prime}_{2p-2}(x_n)} (2p-1)!\, \Delta^{2p} y_{m-p+k} \right| \tag{3.15}$$

$$\leqslant \frac{(2p-1)!}{P^*_{2p-2}(-1)} = \frac{(2p-1)!}{\max\limits_{0\leqslant t\leqslant 1} |Q_{2p-1}(t)|}.$$

Letting $N\to\infty$ in (3.14) and (3.15), we obtain (3.8). For the function to pass through the given points, it is enough to prescribe it at n points, for example at $x = 0, 1, \cdots, n-1$. From (3.7) and (3.8) we obtain

$$\sup_{Y\in\Delta_1^n}\ \inf_{f(x)\in f(x,Y)}\ \bigvee_{-\infty}^{\infty} f^{(n-1)}(x) = \frac{(n-1)!}{\max\limits_{0\leqslant t\leqslant 1}|Q_{n-1}(t)|},$$

where $Q_n(t)$ is defined by (4).

It follows from [1] and the results of the present paper that the fundamental theorem holds for $1\leqq p\leqq\infty$, where $(\int_0^1|Q_n(x)|^q dx)^{1/q}$ is to be interpreted as $\max_{0\leqq x\leqq 1}|Q_n(x)|$ when $q=\infty$.

§4. Functions of several variables

Let n_1, n_2 be fixed positive integers, $n = n_1+n_2$, and let values $y_{k,m}$ $(k, m = 0, \pm 1, \pm 2, \cdots)$ be assigned at the points with integral coordinates; let the nth difference

$$\Delta^{n_1+n_2} y_{k,m} = \sum_{l=0}^{n_1}(-1)^{n_1-l}C_{n_1}^l\sum_{s=0}^{n_2}(-1)^{n_2-s}C_{n_2}^s y_{k+l,\,m+s} \tag{4.1}$$

satisfy

$$|\Delta^{n_1+n_2} y_{k,m}|\leqslant 1\quad (k, m = 0, \pm 1, \pm 2, \ldots). \tag{4.2}$$

We are to find the smallest value, for all sequences $\{y_{k,m}\}$ satisfying this inequality, of the constant A_{n_1,n_2} in the inequality

$$\inf\sup_{x,\,y}\left|\frac{\partial^n f(x,y)}{\partial x^{n_1}\partial y^{n_2}}\right|\leqslant A_{n_1,n_2},$$

where inf is taken over all functions for which $f(k,m) = y_{k,m}$ $(k, m = 0, \pm 1, \cdots)$. We also consider the analogous problem, with appropriate changes in the hypotheses, for $1 < p < \infty$. The results of this section can be extended in a trivial way to functions of a larger number of variables.

We shall look for the required function in the form

$$\begin{aligned} f(x,y) = {} & \frac{1}{(n_1-1)!\,(n_2-1)!}\int_0^x (x-t)^{n_1-1}dt\int_0^y (y-z)^{n_2-1}\frac{\partial^n f(t,z)}{\partial t^{n_1}\partial z^{n_2}}dz \\ & + \sum_{k=0}^{n_1-1}s_k(y)\,x^k + \sum_{k=0}^{n_2-1}\psi_k(x)\,y^k. \end{aligned} \tag{4.3}$$

We consider the values of the function at the integral points and form the differences (4.1). We have

$$\begin{aligned} \Delta^{n_1+n_2}y_{k,m} = {} & \frac{1}{(n_1-1)!\,(n_2-1)!}\int_0^1 dt\int_0^1\sum_{l=0}^{n_1}\sum_{p=l}^{n_1}(-1)^{n_1-p}C_{n_1}^p(p+1-l-t)^{n_1-1} \\ & \times\sum_{s=0}^{n_2}\frac{\partial^n f(t+k+l-1,\, z+m+s-1)}{\partial t^{n_1}\partial z^{n_2}}\sum_{r=s}^{n_2}(-1)^{n_2-r}C_{n_2}^r(r+1-s-z)^{n_2-1}dz. \end{aligned} \tag{4.4}$$

We introduce the following notation.

Y is the sequence $\{y_{k,m}\}$, $k, m = 0, \pm 1, \pm 2, \cdots$.

Δ^{n1+n2} is the class of sequences Y satisfying (4.2).

$F(Y)$ is the class of functions defined in the whole plane and passing through the points $\{y_{k,m}, k, m\}$, $k, m = 0, \pm 1, \pm 2, \cdots$.

We now find a lower bound for A_{n_1,n_2}. We shall show that for the sequence $\tilde{Y} = \{\tilde{y}_{k,m}\}$ for which $\Delta^{n1+n2}\tilde{y}_{k,m} = (-1)^{k+m}$ and for every $f(x,y) \in F(\tilde{Y})$, we have

$$\sup_{x,\, y} \left| \frac{\partial^n f(x, y)}{\partial x^{n_1} \partial y^{n_2}} \right| \geqslant \frac{(n_1 - 1)!\,(n_2 - 1)!}{\int_0^1 |Q_{n_1-1}(t)|\,dt \int_0^1 |Q_{n_2-1}(t)|\,dt}, \tag{4.5}$$

where

$$Q_{n-1}(t) = \sum_{l=0}^{n} \sum_{k=0}^{n-l} (-1)^{n-k} C_n^k \left(k \mp \frac{1}{2} - \frac{t}{2}\right)^{n-1}.$$

Let $f(x,y) \in F(\tilde{Y})$ and

$$\sup_{x,\, y} \left| \frac{\partial^n f(x, y)}{\partial x^{n_1} \partial y^{n_2}} \right| = M(n_1, n_2).$$

We may suppose that $M(n_1, n_2)$ is finite, since in the opposite case (4.5) is certainly true.

For $N_1 > n_1$ and $N_2 > n_2$ we use (4.3) and form the sum

$$\sum_{m=0}^{N_1} (-1)^m \sum_{k=0}^{N_2} (-1)^k \Delta^{n_1+n_2} y_{k,\,m}.$$

By considerations similar to those used in [1] for obtaining a lower bound, we obtain

$$M(n_1, n_2) \geqslant \frac{N_1 \cdot N_2 (n_1 - 1)!\,(n_2 - 1)!}{N_1 N_2 \int_0^1 |Q_{n_1-1}(t)|\,dt \int_0^1 |Q_{n_2-1}(t)|\,dt + O(N_1) + O(N_2)}.$$

Letting N_1 and $N_2 \to \infty$, we obtain (4.5).

We now obtain an upper bound. We shall show that for every sequence $\{y_{k,m}\}$ satisfying (4.2) there is a function $f^*(x,y)$ passing through the points $\{y_{k,m}, k, m\}$ for which

$$\sup_{x,\, y} \left| \frac{\partial^n f^*(x, y)}{\partial x^{n_1} \partial y^{n_2}} \right| \leqslant \frac{(n_1 - 1)!\,(n_2 - 1)!}{\int_0^1 |Q_{n_1-1}(t)|\,dt \int_0^1 |Q_{n_2-1}(t)|\,dt}. \tag{4.6}$$

We make the construction as follows: if $n_1 = 2p_1 + 1$, $n_2 = 2p_2 + 1$, we look for $\partial^n f(x,y)/\partial x^{n1} \partial y^{n2}$ in the form

$$\frac{\partial^n f(x, y)}{\partial x^{n_1}\partial y^{n_2}} = Z_{k,m}, \quad k < x \leqslant k+1, \quad m < y \leqslant m+1; \tag{4.7}$$

if $n_1 = 2p_1$, $n_2 = 2p_2 + 1$, then

$$\frac{\partial^n f(x, y)}{\partial x^{n_1}\partial y^{n_2}} = Z_{k,m}, \quad k - \frac{1}{2} < x \leqslant k + \frac{1}{2}, \quad m < y \leqslant m+1; \tag{4.8}$$

if $n_1 = 2p_1 + 1$, $n_2 = 2p_2$, then

$$\frac{\partial^n f(x, y)}{\partial x^{n_1}\partial y^{n_2}} = Z_{k,m}, \quad k < x \leqslant k+1, \quad m - \frac{1}{2} < y \leqslant m + \frac{1}{2}; \tag{4.9}$$

if $n_1 = 2p_1$, $n_2 = 2p_2$, then

$$\frac{\partial^n f(x, y)}{\partial x^{n_1}\partial y^{n_2}} = Z_{k,m}, \quad k - \frac{1}{2} < x \leqslant k + \frac{1}{2}, \quad m - \frac{1}{2} < y \leqslant m + \frac{1}{2}, \tag{4.10}$$

where $k, m = 0, \pm 1, \pm 2, \cdots$.

Substituting these values into (4.4), we obtain the difference equations

$$(2p_1)!\,(2p_2)!\,\Delta^{n_1+n_2} y_{k,m} = \sum_{l=0}^{2p_1} b_{p_1,l} \sum_{s=0}^{2p_2} b_{p_2,s} Z_{k+l,\,m+s}, \tag{4.11}$$

$$(2p_1 - 1)!\,(2p_2)!\,\Delta^{n_1+n_2} y_{k,m} = \sum_{l=0}^{2p_1} a_{p_1,l} \sum_{s=0}^{2p_2} b_{p_2,s} Z_{k+l,\,m+s}, \tag{4.12}$$

$$(2p_1)!\,(2p_2 - 1)!\,\Delta^{n_1+n_2} y_{k,m} = \sum_{l=0}^{2p_1} b_{p_1,l} \sum_{s=0}^{2p_2} a_{p_2,s} Z_{k+l,\,m+s}, \tag{4.13}$$

$$(2p_1 - 1)!\,(2p_2 - 1)!\,\Delta^{n_1+n_2} y_{k,m} = \sum_{l=0}^{2p_1} a_{p_1,l} \sum_{s=0}^{2p_2} a_{p_2,s} Z_{k+l,\,m+s}, \tag{4.14}$$

where

$$a_{p,l} = \int_{\frac{1}{2}}^{1} \sum_{k=l}^{2p} (-1)^k C_{2p}^k (k - l + 1 - t)^{2p-1}\,dt + \int_0^{\frac{1}{2}} \sum_{k=l+1}^{2p} (-1)^k C_{2p}^k (k - l - t)^{2p-1}\,dt \tag{4.15}$$

and

$$b_{p,l} = -\int_0^1 \sum_{k=l+1}^{2p+1} (-1)^k C_{2p+1}^k (k - l - t)^{2p}\,dt. \tag{4.16}$$

We introduce the notation

$$C_{p,l}^i = \begin{cases} b_{p,l}, & i = 1, \\ a_{p,l}, & i = 2. \end{cases}$$

Then (4.11)—(4.14) can be written in the form

$$(n_1 - 1)!\,(n_2 - 1)!\,\Delta^{n_1+n_2} y_{k,m} = \sum_{l=0}^{2p_1} C_{p_1,l}^j \sum_{s=0}^{2p_2} C_{p_2,s}^i Z_{k+l,\,m+s}, \tag{4.17}$$

where $i, j = 1, 2$.

Let

$$\sum_{s=0}^{2p_2} C^{i}_{p_2, l} Z_{k+l,\, m+s} = V_{k+l,\, m}. \tag{4.18}$$

Then we obtain the difference equation

$$C_{k,\, m} = (n_1 - 1)!\,(n_2 - 1)!\,\Delta^{n_1+n_2} y_{k,\, m} = \sum_{l=0}^{2p_1} C^{j}_{p_1,\, l} V_{k+l,\, m}, \tag{4.19}$$

which for $j = 1, 2$ coincides with the difference equations (27), (29) of [1]. The same is true for (4.18).

Consequently the characteristic polynomial of (4.19) satisfies the hypotheses of Theorem 1 of the Introduction, and the equation has the solution

$$V_{k,\, m} = \sum_{l=-\infty}^{\infty} \sum_{p=1}^{p_1} \frac{x_{p,\, j}^{p_1 - 1 + |l|}}{P'_{2p_1,\, j}(x_{p,\, j})} C_{k-p_1+l,\, m}. \tag{4.20}$$

Moreover,

$$\sup_k |V_{k,\, m}| \leqslant \frac{(-1)^{p_1}}{P_{2p_1,\, j}(-1)} \sup_k |C_{k,\, m}|. \tag{4.21}$$

For (4.18) we have

$$Z_{k+l,\, m} = \sum_{s=-\infty}^{\infty} \sum_{p=1}^{p_2} \frac{x_{p,\, i}^{p_2 - 1 + |s|}}{P'_{2p_2,\, i}(x_{p,\, i})} V_{k+l,\, m-p_2+s}$$

and

$$\sup_m |Z_{k+l,\, m}| \leqslant \frac{(-1)^{p_2}}{P_{2p_2,\, i}(-1)} \sup_m |V_{k+l,\, m}|. \tag{4.22}$$

From (4.21) and (4.22) we obtain

$$\sup_{k,\, m} |Z_{k,\, m}| \leqslant \frac{(-1)^{p_1+p_2}}{P_{2p_1,\, j}(-1)\, P_{2p_2,\, i}(-1)} \sup_{k,\, m} |C_{k,\, m}|. \tag{4.23}$$

From (4.7)—(4.10) and Lemma 5 of [1] we finally obtain

$$\sup_{x,\, y} \left| \frac{\partial^n f(x, y)}{\partial x^{n_1} \partial y^{n_2}} \right| \leqslant \frac{(n_1 - 1)!\,(n_2 - 1)!}{\int_0^1 |Q_{n_1-1}(t)|\, dt \int_0^1 |Q_{n_2-1}(t)|\, dt} \tag{4.24}$$

The function so obtained has the same differences $\Delta^{n_1+n_2} f(k, m)$ as the given sequence. Requiring that $f(k, m) = y_{k,m}$ for $k = 0, 1, \cdots, n_1 - 1$, $m = 0, \pm 1, \pm 2, \cdots$, and $f(k, m) = y_{k,m}$ for $m = 0, 1, \cdots, n_2 - 1$, $k = 0, \pm 1, \pm 2, \cdots$ we obtain a function $f^*(x, y)$ that belongs to $F(Y)$ and satisfies (4.6).

From (4.6) and (4.5) we obtain

$$\sup_{Y \in \Delta^{n_1+n_2}} \inf_{f(x, y) \in F(y)} \sup_{x, y} \left| \frac{\partial^n f(x, y)}{\partial x^{n_1} \partial y^{n_2}} \right| = \frac{(n_1-1)!\,(n_2-1)!}{\int_0^1 |Q_{n_1-1}(t)|\,dt \int_0^1 |Q_{n_2-1}(t)|\,dt}.$$

It follows from the results obtained here and in [1] that $A_{n1,n2} = A_{n1}A_{n2}$.

We now consider the analogous problem in $L_p^{(2)}$, where $L_p^{(2)}$ is the space of functions defined over the space of two real variables and integrable together with their pth powers.

Let n_1 and n_2 be fixed positive integers, $n = n_1 + n_2$; let the values $y_{k,m}$, $k, m = 0, \pm 1, \pm 2, \cdots$, be given at the points with integral coordinates; and let the nth differences defined by (4.1) satisfy

$$\left(\sum_{k=-\infty}^{\infty} \sum_{m=-\infty}^{\infty} |\Delta^{n_1+n_2} y_{k,m}|^p \right)^{\frac{1}{p}} \leqslant 1 \quad (1 < p < \infty). \tag{4.25}$$

We are to find the smallest value, for all sequences $\{y_{k,m}\}$ satisfying (4.25) of the constant $A_{n1,n2,p}$ in the inequality

$$\inf \left(\int_{-\infty}^{\infty} dy \int_{-\infty}^{\infty} \left| \frac{\partial^n f(x, y)}{\partial x^{n_1} \partial y^{n_2}} \right|^p dx \right)^{\frac{1}{p}} \leqslant A_{n_1, n_2, p}, \tag{4.26}$$

where inf is taken over all functions for which $f(k, m) = y_{k,m}$ $(k, m = 0, \pm 1, \cdots)$ and $\{y_{k,m}\}$ belongs to $\Delta_p^{n_1+n_2}$, where $\Delta_p^{n_1+n_2}$ is the class of sequences $Y = \{y_{k,m}\}$ satisfying (4.25).

We first obtain a lower bound for $A_{n1,n2,p}$. In fact, we show that

$$A_{n_1, n_2, p} \geqslant \frac{(n_1-1)!\,(n_2-1)!}{\left(\int_0^1 |Q_{n_1-1}(t)|^q dt \right)^{\frac{1}{q}} \left(\int_0^1 |Q_{n_2-1}(t)|^q dt \right)^{\frac{1}{q}}}, \quad \left(\frac{1}{p} + \frac{1}{q} = 1 \right). \tag{4.27}$$

Let

$$a_{n_1, l}(t) = \sum_{r=l}^{n_1} (-1)^{n_1-r} C_{n_1}^r (r + 1 - l - t)^{n_1-1}.$$

To obtain (4.27), we use (4.4) and the notation introduced above; we consider the sequence $\widetilde{Y} \in \Delta_p^{n1+n2}$ for which

$$\Delta^{n_1+n_2} \widetilde{y}_{k, m} = \frac{(-1)^{k+m}}{(2N+1)^{\frac{2}{p}}} \quad \text{for } \max(|k|, |m|) \leqslant N$$

and

$$\Delta^{n_1+n_2} \widetilde{y}_{k, m} = 0 \quad \text{for } \max(|k|, |m|) > N,$$

and form the sum

$$\sum_{k=-N}^{N}\sum_{m=-N}^{N}(-1)^{k+m}\Delta^{n_1+n_2}\tilde{y}_{k,m}=\sum_{k=-N}^{N}\sum_{m=-N}^{N}\frac{1}{(2N+1)^{\frac{2}{p}}}=(2N+1)^{\frac{2}{q}} \tag{4.28}$$

$$=\sum_{k=-N}^{N}\sum_{m=-N}^{N}\frac{(-1)^{k+m}}{(n_1-1)!\,(n_2-1)!}\int_0^1 dt\int_0^1\sum_{l=0}^{n_1}a_{n_1,l}(t)\sum_{s=0}^{n_2}a_{n_2,s}(z)$$
$$\times\frac{\partial^n f(t+k+l-1,\,z+m+s-1)}{\partial t^{n_1}\partial z^{n_2}}\,dz.$$

We define $Q_n(x)$ by (4), and put

$$s_{k,m}(t,z)=\frac{\partial^n f(t+k-1,\,z+m-1)}{\partial t^{n_1}\partial z^{n_2}}.$$

We may suppose that there is a function $f(x,y)\in F(\tilde{Y})$ for which

$$M_n=\Big(\int_{-\infty}^{\infty}dy\int_{-\infty}^{\infty}\Big|\frac{\partial^n f(x,y)}{\partial n^{n_1}\partial y^{n_2}}\Big|^p dx\Big)^{\frac{1}{p}}$$

is finite. In the contrary case (4.27) holds automatically.

From (4.28) we have

$$(n_1-1)!\,(n_2-1)!\,(2N+1)^{\frac{2}{q}}\leqslant\sum_{k=-N}^{N}\int_0^1|Q^*_{n_1-1}(t)|\,dt\sum_{m=-N}^{N}\int_0^1|Q^*_{n_2-1}(z)|$$
$$\times|s_{k,m}(t,z)|\,dz$$
$$+\sum_{k=-N}^{N}\int_0^1|Q^*_{n_1-1}(t)|\,dt\Big[\sum_{m=N+1}^{N+n_2}\int_0^1|s_{k,m}(t,z)|\sum_{s=m-N}^{n_2}|a_{n_2,s}(z)|\,dz$$
$$+\sum_{m=-N}^{-N+n_2-1}\int_0^1|s_{k,m}(t,z)|\sum_{s=N+m+1}^{n_2}|a_{n_2,s}(z)|\,dz\Big]$$
$$+\sum_{k=N+1}^{N+n_1}\int_0^1 dt\sum_{l=m-N}^{n_1}|a_{n_1,l}(t)|\Big[\sum_{m=-N}^{N}\int_0^1|Q^*_{n_2-1}(z)|\,|s_{k,m}(t,z)|\,dz$$
$$+\sum_{m=N+1}^{N+n_2}\int_0^1\sum_{s=m-N}^{n_2}|a_{n_2,s}(z)|\,|s_{k,m}(t,z)|\,dz$$
$$+\sum_{m=-N}^{-N+n_2-1}\int_0^1\sum_{s=N+m+1}^{n_2}|a_{n_2,s}(z)|\,|s_{k,m}(t,z)|\,dz\Big]$$
$$+\sum_{k=-N}^{-N+n-1}\int_0^1 dt\sum_{l=N+m-1}^{n_1}|a_{n_1,l}(t)|\Big[\sum_{m=-N}^{N}\int_0^1|Q^*_{n_2-1}(z)|\,|s_{k,m}(t,z)|\,dz$$
$$+\sum_{m=N+1}^{N+n_2}\int_0^1\sum_{s=m-N}^{n_2}|a_{n_2,s}(z)|\,|s_{k,m}(t,z)|\,dz$$
$$+\sum_{m=-N}^{-N+n_2-1}\int_0^1\sum_{s=m+N-1}^{n_2}|a_{n_2,s}(z)|\,|s_{k,m}(t,z)|\,dz\Big].$$

We put $s_{k,m}(t,z)=\partial^n f(t+k-1,\ z+m-1)/\partial t^{n_1}\partial z^{n_2}$ and define $Q_n(x)$ by (4). We may suppose there is a function $f(x,y)\in F(\widetilde{Y})$ for which

$$\sum_{m=-N}^{N}\int_0^1 |Q^*_{n_2-1}(z)|\,dz\sum_{k=-N}^{N}\int_0^1 |Q^*_{n_1-1}(t)|\,|s_{k,m}(t,z)|\,dt$$

$$\leqslant\sum_{m=-N}^{N}\int_0^1 |Q^*_{n_2-1}(z)|\,dz\sum_{k=-N}^{N}\left(\int_0^1 |Q^*_{n_1-1}(t)|^q\,dt\right)^{\frac{1}{q}}\left(\int_0^1 |s_{k,m}(t,z)|^p dt\right)^{\frac{1}{p}}$$

$$\leqslant(2N+1)^{\frac{1}{q}}\left(\int_0^1 |Q^*_{n_1-1}(t)|^q dt\right)^{\frac{1}{q}}\sum_{m=-N}^{N}\int_0^1 |Q^*_{n_2-1}(z)|\,dz\left(\sum_{k=-N}^{N}\int_0^1 |s_{k,m}(t,z)|^p dt\right)^{\frac{1}{p}}$$

$$\leqslant(2N+1)^{\frac{1}{q}}\left(\int_0^1 dt\int_0^1 |Q^*_{n_1-1}(t)\,Q^*_{n_2-1}(z)|^q dt\right)^{\frac{1}{q}}\sum_{m=-N}^{N}\left[\int_0^1 dz\left(\sum_{k=-N}^{N}|s_{k,m}(t,z)|^p dz\right)\right]^{\frac{1}{p}}$$

$$\leqslant(2N+1)^{\frac{2}{q}}\left(\int_0^1 dt\int_0^1 |Q^*_{n_1-1}(t)\,Q^*_{n_2-1}(z)|^q dz\right)^{\frac{1}{q}}\left[\sum_{m=-N}^{N}\int_0^1 dz\sum_{k=-N}^{N}\int_0^1 |s_{k,m}(t,z)|^p dz\right]^{\frac{1}{p}}.$$

Treating the other terms similarly, replacing t by $(1+t)/2$ and z by $(1+z)/2$ and using the fact that $M_{n1,n2,p}\leqq A_{n1,n2,p}$ we obtain for all N, the inequality

$$A_{n_1,n_2,p}\geqslant M_n\geqslant\frac{(n_1-1)!\,(n_2-1)!\,(2N+1)^{\frac{2}{q}}}{(2N+1)^{\frac{2}{q}}\left(\int_0^1 |Q_{n_1-1}(t)|^q dt\right)^{\frac{1}{q}}\left(\int_0^1 |Q_{n_2-1}(t)|^q dt\right)^{\frac{1}{q}}+O\left[(2N+1)^{\frac{1}{q}}\right]}.$$

Letting $N\to\infty$, we obtain (4.27).

We now obtain an upper bound. We show that for every $\{y_{k,m}\}$ satisfying (4.25) there is a function $f^*(x,y)$ passing through the points $\{k,m,y_{k,m}\}$ for which

$$\left(\int_{-\infty}^{\infty}dx\int_{-\infty}^{\infty}\left|\frac{\partial^n f(x,y)}{\partial x^{n_1}\partial y^{n_2}}\right|^p dy\right)^{\frac{1}{p}}\leqslant\frac{(n_1-1)!\,(n_2-1)!}{\left(\int_0^1 |Q_{n_1-1}(t)|^q dt\right)^{\frac{1}{q}}\left(\int_0^1 |Q_{n_2-1}(t)|^q dt\right)^{\frac{1}{q}}}. \tag{4.29}$$

We make the construction as follows: if $n_1=2p_1+1$, $n_2=2s+1$, we look for $\partial^n f(x,y)/\partial x^{n1}\partial y^{n2}$ in the form

$$\frac{\partial^n f(k+t,\ m+z)}{\partial t^{2p_1+1}\partial z^{2s+1}}=Z_{m+1,\,k+1}\,|Q^*_{2p_1}(t)\,Q^*_{2s}(z)|^{q-1}, \tag{4.30}$$

where $0<t\leqq 1$, $0<z\leqq 1$, $k,m=0,\pm 1,\pm 2,\cdots$; if $n_1=2p_1+1$, $n_2=2s$, then

$$\frac{\partial^n f(t+k,\ m+z)}{\partial t^{2p_1+1}\partial z^{2s}}=\begin{cases}Z_{m+1,\,k}\,|Q^*_{2p_1}(t)\,Q^*_{2s-1}(1+z)|^{q-1}, & 0<t\leqslant 1;\ -\frac{1}{2}<z\leqslant 0\\ Z_{m+1,\,k}\,|Q^*_{2p_1}(t)\,Q^*_{2s-1}(z)|^{q-1}, & 0<t\leqslant 1;\ 0<z\leqslant\frac{1}{2};\end{cases} \tag{4.31}$$

if $n_1 = 2p_1$, $n_2 = 2s+1$, then

$$\frac{\partial^n f(k+t, m+z)}{\partial t^{2p_1}\partial z^{2s+1}} = \begin{cases} Z_{m,k+1}\,|\,Q^*_{2p_1-1}(1+t)\,Q^*_{2s}(z)\,|^{q-1}, & -\frac{1}{2}<t\leqslant 0;\ 0<z\leqslant 1, \\ Z_{m,k+1}\,|\,Q^*_{2p_1-1}(t)\,Q^*_{2s}(z)\,|^{q-1}, & 0<t\leqslant\frac{1}{2};\ 0<z\leqslant 1; \end{cases} \tag{4.32}$$

if $n_1 = 2p_1$, $n_2 = 2p_2$, then

$$= \begin{cases} Z_{k,m}\,|\,Q^*_{2p_1-1}(1+t)\,Q^*_{2s-1}(1+z)\,|^{q-1}, & -\frac{1}{2}<t\leqslant 0, \quad -\frac{1}{2}<z\leqslant 0; \\ Z_{k,m}\,|\,Q^*_{2p_1-1}(1+t)\,Q^*_{2s-1}(z)\,|^{q-1}, & -\frac{1}{2}<t\leqslant 0, \ 0<z\leqslant\frac{1}{2}; \\ Z_{k,m}\,|\,Q^*_{2p_1-1}(t)\,Q^*_{2s-1}(1+z)\,|^{q-1}, & 0<t\leqslant\frac{1}{2}, \quad -\frac{1}{2}<z\leqslant 0; \\ Z_{k,m}\,|\,Q^*_{2p_1-1}(t)\,Q^*_{2s-1}(z)\,|^{q-1}, & 0<t\leqslant\frac{1}{2}, \quad 0<z\leqslant\frac{1}{2}; \end{cases} \tag{4.33}$$

$k, m = 0, \pm 1, \pm 2, \cdots$.

Substituting (4.30)—(4.33) into (4.4), we obtain the difference equation

$$\sum_{l=0}^{2p_1} C^{(p_1)}_{l,i} \sum_{r=0}^{2s} C^{(s)}_{r,j} Z_{k+l,\, m+r} = (n_1-1)!\,(n_2-1)!\,\Delta^{n_1+n_2} y_{k,m}, \tag{4.34}$$

where $C^{(p1)}_{l,1} = a_{p1,l}$, $C^{(p1)}_{l,2} = b_{p1,l}$ and

$$a_{p_1,l} = \int_0^1 |\,Q^*_{2p_1}(t)\,|^{q-1} \sum_{k=l}^{2p_1+1} (-1)^{k-1} C^k_{2p+1} (k+1-l-t)^{2p}\,dt, \tag{4.35}$$

$$b_{p_1,l} = \int_{\frac{1}{2}}^{1} |\,Q^*_{2p_1-1}(t)\,|^{q-1} \sum_{k=l}^{2p_1} (-1)^k C^k_{2p_1} (k-l+1-t)^{2p_1-1}\,dt$$

$$+ \int_0^{\frac{1}{2}} |\,Q^*_{2p_1-1}(t)\,|^{q-1} \sum_{k=l+1}^{2p_1} (-1)^k C^k_{2p_1} (k-l+1-t)^{2p_1-1}\,dt. \tag{4.36}$$

Equation (4.34) is similar to (4.17). Hence we can solve (4.34) in the same way as we solved (4.17), since its characteristic polynomial satisfies the hypotheses of Theorem 1.

The last statement follows from the results of §2 of the present paper.

In the same way as in §1 we can show that

$$\left\| \frac{\partial^n f(x, y)}{\partial x^{n_1} \partial y^{n_2}} \right\|_{L_p^{(2)}} \leqslant \frac{(n-1)!\,(n_2-1)!}{\left(\int_0^1 |Q_{n_1-1}(t)|^q dt\right)^{\frac{1}{q}} \left(\int_0^1 |Q_{n_2-1}(t)|^q dt\right)^{\frac{1}{q}}}, \tag{4.37}$$

where $\partial^n f(x,y)/\partial x^{n_1}\partial y^{n_2}$ is determined by (4.30)—(4.33) and the difference equation (4.34).

Requiring that the function passes through the given points does not affect $\Delta^{n_1+n_2} f(k,m)$. Hence we have obtained a function passing through the points of the given sequence and satisfying (4.37). From (4.37) and (4.27) we obtain

$$\sup_{Y \in \Delta_p^{n_1+n_2}} \inf_{f(x, y) \in F(Y)} \left(\int_{-\infty}^{\infty} dx \int_{-\infty}^{\infty} \left| \frac{\partial^n f(x, y)}{\partial x^{n_1} \partial y^{n_2}} \right|^p dy \right)^{\frac{1}{p}} = A_{n_1, n_2, p}, \tag{4.38}$$

where

$$A_{n_1, n_2, p} = \frac{(n_1-1)!\,(n_2-1)!}{\left(\int_0^1 |Q_{n_1-1}(t)|^q dt\right)^{\frac{1}{q}} \left(\int_0^1 |Q_{n_2-1}(t)|^q dt\right)^{\frac{1}{q}}}, \quad \frac{1}{p}+\frac{1}{q}=1, \quad p>1.$$

It follows from the results of this section and §2 that

$$A_{n_1, n_2, p} = A_{n_1, p} A_{n_2, p}.$$

Let the differences defined by (4.1) satisfy

$$\left[\sum_{k=-\infty}^{\infty} \left(\sum_{m=-\infty}^{\infty} |\Delta^n y_{k, m}|^{p_1} \right)^{\frac{p_2}{p_1}} \right]^{\frac{1}{p_2}} \leqslant 1 \quad (1 < p_1 \leqslant p_2 \leqslant \infty). \tag{4.39}$$

We write $Y \in \Delta_{p_1+p_2}^{n_1+n_2}$ if the sequence Y satisfies (4.39). Then we have

Theorem 4.

$$\sup_{Y \in \Delta_{p_1, p_2}^{n_1+n_2}} \inf_{f(x, y) \in F(Y)} \left[\int_{-\infty}^{\infty} dy \left(\int_{-\infty}^{\infty} \left| \frac{\partial^n f(x, y)}{\partial x^{n_1} \partial y^{n_2}} \right|^{p_1} dx \right)^{\frac{p_2}{p_1}} \right]^{\frac{1}{p_2}} = A_{n_1, n_2, p_1, p_2}, \tag{4.40}$$

where

$$A_{n_1, n_2, p_1, p_2} = \frac{(n_1-1)!\,(n_2-1)!}{\left(\int_0^1 |Q_{n_1-1}(t)|^{q_1} dt\right)^{\frac{1}{q_1}} \left(\int_0^1 |Q_{n_2-1}(t)|^{q_2} dt\right)^{\frac{1}{q_2}}} \tag{4.41}$$

$$\left(\frac{1}{p_1}+\frac{1}{q_1}=1, \quad \frac{1}{p_2}+\frac{1}{q_2}=1\right).$$

Equation (4.40) is proved by reasoning similar to that used for (4.32). Comparing (4.41) and (7), we obtain

$$A_{n_1, n_2, p_1, p_2} = A_{n_1, p_1} A_{n_2, p_2}.$$

Bibliography

[1] Ju. N. Subbotin, *On the relations between finite differences and the corresponding derivatives,* Trudy Mat. Inst. Steklov. **78** (1965), 24-42 = Proc. Steklov. Inst. Math. **78** (1965), 23-42. MR **32** # 7978.

[2] N. I. Ahiezer, *Lectures in the theory of approximation,* OGIZ, Moscow, 1947; 2nd rev. ed., "Nauka", Moscow, 1965; English transl., Ungar, New York, 1956. MR **10**, 33; MR **20** # 1872; MR **32** # 6108.

[3] S. K. Godunov and V. S. Rjaben'kiĭ, *Canonical forms of systems of linear ordinary difference equations with constant coefficients,* Ž. Vyčisl. Mat. i Mat. Fiz. **3** (1963), 211-222. (Russian) MR **27** # 5054.

[4] V. S. Rjaben'kiĭ, *Necessary and sufficient conditions for best conditionality of boundary-value problems for systems of ordinary difference equations,* Ž. Vyčisl. Mat. i Mat. Fiz. **4** (1964), 242-255. (Russian) MR **28** # 5574.

[5] ———, *On the stability of finite difference schemes and on the application of the finite difference method to the solution of the Cauchy problem for systems of partial differential equations,* Dissertation, Moskov. Gos. Univ., 1952.

[6] V. S. Rjaben'kiĭ and A. F. Filippov. *On the stability of difference equations,* GITTL, Moscow, 1956, pp. 158-166; German transl., Mathematik für Naturwiss. und Technik, Band 3, VEB Deutscher Verlag, Berlin, 1960. MR **19**, 865; MR **23** # 437.

[7] S. L. Sobolov, *Lectures on the theory of cubature formulas.* Part II, Novosibirsk, 1965, p. 170. (Russian) MR **35** # 3884.

[8] A. O. Gel'fond, *Calculus of finite differences,* Fizmatgiz, Moscow, 1959; French transl. Collection Univ. de Math., XII, Dunod, Paris, 1963. MR **35** # 7020; MR **28** # 376.

[9] G. M. Fihtengol'c, *A course of differential and integral calculus.* Vol. 1, Fizmatgiz, Moscow, 1958; German transl., Hochschulbücher für Mathematik, Band 61, VEB Deutscher Verlag, Berlin, 1964, 1966.

[10] A. Peyerimhoff, *On convergence fields of Nörlund means,* Proc. Amer. Math. Soc. 7 (1956), 335-347. MR **17**, 1199.

ON APPROXIMATING SOME CLASSES OF PERIODIC FUNCTIONS IN MEAN

L. V. TAĬKOV

Let r be a positive integer. In the present paper we consider real functions of period 2π whose $(r-1)$th derivatives are absolutely continuous. Let

$$S_n(f, x) = \frac{a_0}{2} + \sum_{k=1}^{n} (a_k \cos kx + b_k \sin kx)$$

be the partial sums of order n of the Fourier series of $f(x)$.

In 1936 J. Favard obtained an estimate for the deviation of a function from the partial sums of its Fourier series in the metric of the space C of continuous functions, in terms of the corresponding deviation for the rth derivative:

$$\|f - S_{n-1}(f)\|_C \leqslant K_r n^{-r} \|f^{(r)} - S_{n-1}(f^{(r)})\|_C, \quad n \geqslant 1. \tag{1}$$

This becomes an equality for $f(x) = \varphi_{r,n}(x) = n^{-r}\varphi_r(nx)$, where

$$\varphi_r(x) = \frac{4}{\pi} \sum_{k=1}^{\infty} \frac{\sin\left\{(2k-1)x - \frac{\pi r}{2}\right\}}{(2k-1)^{r+1}}.$$

Put

$$T_n(x) = \frac{\alpha_0}{2} + \sum_{k=1}^{n} (\alpha_k \cos kx + \beta_k \sin kx)$$

and let

$$E_n(f, C) = \inf_{T_n} \|f - T_n\|_C$$

be the best approximation of $f(x)$ by trigonometric polynomials $T_n(x)$ of order at most n in the C metric. As Sun Yung-sheng [2] remarked, it can be shown in the same way that

$$E_{n-1}(f, C) \leqslant K_r n^{-r} E_{n-1}(f^{(r)}, C). \tag{2}$$

Here there is also equality for $\varphi_{r,n}(x)$. Inequality (2), with a constant twice as large, had already been proved by S. B. Stečkin [3] as a lemma.

There are many generalizations of (1) and (2) in the C metric. We are interested in analogues of these inequalities in the metrics of Orlicz spaces.

In this connection it is appropriate to recall some definitions (see for example [4]).

A function $\Phi(u)$, $u \geqq 0$, is called an N-function if it admits the representation

$$\Phi(u)=\int_0^u p(t)\,dt,$$

where $p(t)$ is a nondecreasing function, continuous on the right, satisfying

$$p(t)>0\ (t>0),\ p(0)=0,\ \lim_{t\to\infty} p(t)=\infty.$$

Since

$$q(s)=\sup_{p(t)\leqslant s} t\ \ (s\geqslant 0)$$

satisfies the same conditions as $p(t)$,

$$\Psi(u)=\int_0^u q(s)\,ds\ \ (u\geqslant 0)$$

is an N-function. $\Phi(u)$ and $\Psi(u)$ are called complementary N-functions. The Orlicz class L_Ψ is the class of Lebesgue measurable functions $v(x)$ of period 2π for which

$$\rho(v,\Psi)=\int_0^{2\pi}\Psi(|v(x)|)\,dx<\infty.$$

The Orlicz space L_Φ^* consists of the Lebesgue measurable functions $u(x)$ of period 2π of finite norm

$$\|u\|_\Phi=\sup_{\rho(v,\Psi)\leqslant 1}\int_0^{2\pi} u(x)\,v(x)\,dx.$$

In the Orlicz space L_Φ^* we can introduce the equivalent Luxemburg norm

$$\|u\|_{(\Phi)}=\inf_{\rho\left(\frac{u}{k},\Phi\right)\leqslant 1} k.$$

We then speak of the Orlicz space with the Luxemburg norm.

The main results of this paper are stated in Theorems 2 and 3; in brief, they consist of the two equations

$$\sup_{\|f^{(r)}-S_{n-1}(f^{(r)})\|_C\leqslant 1}\|f-S_{n-1}(f)\|_{(\Phi)}=\|\varphi_r\|_{(\Psi)}n^{-r}, \tag{1'}$$

$$\sup_{E_{n-1}(f^{(r)},L_\Phi^*)\leqslant 1} E_{n-1}(f,L_1)=\|\varphi_r\|_{(\Psi)}n^{-r}. \tag{2'}$$

How are these equations connected with known results? If we write (1) in the equivalent form

$$\sup_{\|f^{(r)}-S_{n-1}(f^{(r)})\|_C\leqslant 1}\|f-S_{n-1}(f)\|_C=K_r n^{-r},$$

then if we put $\Phi(u) = u^p$ in (1′) and let $p \to \infty$ we obtain this form of (1). On the other hand, if we put $\Phi(u) = u^p$ in (2′) and let $p \downarrow 1$, we obtain

$$\sup_{E_{n-1}(f^{(r)}, L_1) \leqslant 1} E_{n-1}(f, L_1) = K_r n^{-r}.$$

This is close to a result of S. M. Nikol′skiĭ [5], who proved that

$$\sup_{\|f^{(r)}\|_{L_1} \leqslant 1} E_{n-1}(f, L_1) = K_r n^{-r}.$$

A central place in our work is occupied by Theorem 1, in whose proof we use Favard's inequality (1) and a result of A. N. Kolmogorov [6] which we quote below.

Following Kolmogorov, we say that a function $P_r(x) = a\varphi_r(bx + c)$ is an upper comparison function of order $r \geqq 1$ for $f(x)$ at x_0 if

$$\|P_r\|_C \geqslant \|f\|_C, \ \|P_r^{(r)}\|_C = \|f^{(r)}\|_C, \ |P_r(x_0)| = |f(x_0)|.$$

KOLMOGOROV'S THEOREM. *If $P_r(x)$ is an upper comparison function of order r for $f(x)$ at x_0, then*

$$|f'(x_0)| \leqslant |P_r'(x_0)|.$$

In geometric terms this says that the tangents to the graph of the upper comparison function are steeper at the level $y = l$ than the tangents to the graph of $f(x)$. It is in this form that we shall use Kolmogorov's theorem.

THEOREM 1. *Let $\Phi(u)$ be an N-function. Then*

$$\sup_{\|f^{(r)} - S_{n-1}(f^{(r)})\|_C \leqslant 1} \int_0^{2\pi} \Phi(|f(x) - S_{n-1}(f, x)|)\, dx = \int_0^{2\pi} \Phi(n^{-r} |\varphi_r(x)|)\, dx$$

for all positive integers r and n.

PROOF. Since $\varphi_{r,n}^{(r)}(x) = \operatorname{sign} \sin nx$ we have

$$\int_0^{2\pi} \Phi(|\varphi_{r,n}(x) - S_{n-1}(\varphi_{r,n}(x)|)\, dx = \int_0^{2\pi} \Phi(|\varphi_{r,n}(x)|)\, dx;$$

it is then enough to show that

$$\int_0^{2\pi} \Phi(|R(x)|)\, dx \leqslant \int_0^{2\pi} \Phi(|\varphi_{r,n}(x)|)\, dx \tag{3}$$

for every function $R(x) = f(x) - S_{n-1}(f, x)$ with $\|R^{(r)}\|_C = 1$. We may suppose $R(0) = 0$ since otherwise we can replace $R(x)$ by a translation $R(x + \tau)$.

Let $\Delta_k = [a_k, b_k] \subset [0, 2\pi]\ (1 \leqq k \leqq m)$ be the intervals satisfying the following three conditions: the length of each interval is at least πn^{-1}; the interval (a_k, b_k) contains no zeros of $R(x)$; and $R(x)$ vanishes at the end points a_k, b_k, i.e. $R(a_k) = R(b_k) = 0$. The complementary set

$$[0,2\pi] \setminus \bigcup_{k=1}^{m} \Delta_k$$

either is empty or consists of a set of disjoint intervals $\{\delta_k\}$ $(1 \leqq k \leqq l \leqq m)$. If the interval δ_k does not contain a zero of $R(x)$, its length is less than πn^{-1}. If δ_k contains a zero of $R(x)$, then each such zero has, either to its right or its left, another zero of $R(x)$ at a distance less than πn^{-1}. These properties will be used later.

We also note some properties of $\varphi_{r,n}(x)$, $r = 1, 2, \cdots$, which are easily established by induction on r. If ξ is a zero of $\varphi_{r,n}(x)$, this function is monotonic on the interval $[\xi - \pi(2n)^{-1}, \xi + \pi(2n)^{-1}]$ and is odd with respect to ξ. If η is a minimum or maximum of $\varphi_{r,n}(x)$, the function is odd with respect to η. Finally, for each $t \geqq 0$ the set $\bar{e} = \{x: |\varphi'_{r+1,n}(x)| \geqq t\} \cap [\xi - \pi(2n)^{-1}, \xi + \pi(2n)^{-1}]$ is either empty or an interval.

We first show that

$$\int_{\delta_k} \Phi(|R(x)|)\,dx \leqslant \int_{\delta_k} \Phi(|\varphi_{r,n}(x)|)\,dx$$

for each interval δ_k. It is enough to show that

$$\int_{x_1}^{x_2} \Phi(|R(x)|)\,dx \leqslant \int_{x_1}^{x_2} \Phi(|\varphi_{r,n}(x)|)\,dx, \tag{4}$$

where x_1 and x_2 are any points such that $0 \leqq x_1 < x_2$, $R(x_1) = R(x_2)$, $x_2 - x_1 < \pi n^{-1}$. Suppose for the moment that $\varphi_{r,n}(\xi) = 0$ and $\xi = 2^{-1}(x_1 + x_2)$. In this case it is enough to prove the two inequalities

$$\int_{x_1}^{\xi} \Phi(|R(x)|)\,dx \leqslant \int_{x_1}^{\xi} \Phi(|\varphi_{r,n}(x)|)\,dx,$$

$$\int_{\xi}^{x_2} \Phi(|R(x)|)\,dx \leqslant \int_{\xi}^{x_2} \Phi(|\varphi_{r,n}(x)|)\,dx.$$

Since $\varphi_{r,n}(\xi + t) = -\varphi_{r,n}(\xi - t)$, the right-hand sides of these inequalities are equal. We shall establish the first inequality; the second is established similarly.

We have

$$\int_{0}^{2\pi} \Phi(|R(x)|)\,dx \leqslant \int_{0}^{2\pi} \Phi(|\varphi_{r,n}(x)|)\,dx.$$

On (x_1, ξ) we have $|R(x)| \leqq |\varphi_{r,n}(\xi + x_1 - 1)|$. If this were not true, we should have $|R(\eta)| > |\varphi_{r,n}(\xi + x_1 - \eta)|$ at some point $\eta \in (x_1, \xi)$. Let η_1 be the point closest to η on the left such that $x_1 \leqq \eta_1 < \eta$, $|R(\eta_1)| = |\varphi_{r,n}(\xi + x_1 - \eta_1)|$. Then there are points η_2 and η_3 such that $\eta_1 \leqq \eta_2 \leqq \eta_3 < \eta$, $|R(\eta_2)| = |\varphi_{r,n}(\xi + x_1 - \eta_3)|$, and $|R'(\eta_2)| > |\varphi'_{r,n}(\xi + x_1 - \eta_3)|$. The last inequality contradicts Kolmogorov's theorem, since $\varphi_{r,n}(\xi + x_1 - x)$ is an upper comparison function for $R(x)$. In fact, $\|R^{(r)}\|_C = \|\varphi^{(r)}_{r,n}\|_C = 1$; by Favard's inequality $\|R\|_C \leqq \|\varphi_{r,n}\|_C$; finally $|R(\eta_2)| = |\varphi_{r,n}(\xi + x_1 - \eta_3)|$.

We have proved (4) under the assumption that $\varphi_{r,n}((x_1 + x_2)/2) = 0$. However, it is easily seen that

$$\int_{x_1}^{x_2} \Phi(|\varphi_{r,n}(x)|)\,dx \leqslant \int_{x_1}^{x_2} \Phi(|\varphi_{r,n}(x+\tau)|)\,dx,$$

and hence (4) holds in general.

Hence we have

$$\sum_{k=1}^{l} \int_{\delta_k} \Phi(|R(x)|)\,dx \leqslant \sum_{k=1}^{l} \int_{\delta_k} \Phi(|\varphi_{r,n}(x)|)\,dx. \tag{5}$$

We now prove that (for $k = 1, 2, \cdots, m$)

$$\int_{\Delta_k} \Phi(|R(x)|)\,dx \leqslant \int_{\Delta_k} \Phi(|\varphi_{r,n}(x)|)\,dx \tag{6}$$

From the properties stated above for the functions $\varphi_{r,n}(x) = \varphi'_{r+1,n}(x)$, it follows in particular that

$$\int_{\bar{e}} |\varphi'_{r+1,n}(x)|\,dx = 2|\varphi_{r+1,n}(\bar{x})|, \quad |\varphi'_{r+1,n}(\bar{x})| = t,$$

i.e. the variation of $\varphi_{r+1,n}(x)$ on $\bar{e}$ is geometrically equal to the length of the interval $[-|\varphi_{r+1,n}(\bar{x})|, |\varphi_{r+1,n}(\bar{x})|]$ on the axis of ordinates. Since $|\varphi'_{r+1,n}(x)|$ has period πn^{-1} we have

$$\int_{\bar{e}} |\varphi'_{r+1,n}(x)|\,dx = \int_{\bar{e}_\tau} |\varphi_{r+1,n}(x+\tau)|\,dx = 2|\varphi_{r+1,n}(\bar{x})|$$

for every real τ, where $\bar{e}_\tau = \{x : |\varphi'_{r+1,n}(x+\tau)| \geqq t\} \cap [\xi - \pi(2n)^{-1}, \xi + \pi(2n)^{-1}]$, i.e. for every interval $\Delta \subset [0, 2\pi]$ of length $m\Delta = \pi n^{-1}$ the variation of $\varphi_{r+1,n}(x)$ on the part of Δ where $|\varphi'_{r+1,n}(x)| \geqq t$ is equal to $2|\varphi_{r+1,n}(x)|$.

We can now complete the proof of Theorem 1. Let $\rho(x)$ be a function orthogonal to $2n - 1$ harmonics, and let $\rho'(x) = R(x)$. We have to prove

$$\int_{\Delta_k} \Phi(|\rho'(x)|)\,dx \leqslant \int_{\Delta_k} \Phi(|\varphi'_{r+1,n}(x)|)\,dx. \tag{7}$$

Since $\Phi(u)$ is an N-function, $u^{-1}\Phi(u)$ is strictly increasing for $u > 0$ and $\lim u^{-1}\Phi(u) = 0$, $u \downarrow 0$ (see for example [4], p. 18). We also have the identity

$$\Phi(|\rho'(x)|) = \int_0^{|\rho'(x)|} |\rho'(x)|\, d\left\{\frac{\Phi(t)}{t}\right\};$$

we integrate this with respect to x and then change the order of integration. We then obtain

$$\int_{\Delta_k} \Phi(|\rho'(x)|)\, dx = \int_{\Delta_k} \left[\int_0^{|\rho'(x)|} |\rho'(x)|\, d\left\{\frac{\Phi(t)}{t}\right\}\right] dx = \int_0^{\|\rho'\|_{\Delta_k}} d\left\{\frac{\Phi(t)}{t}\right\} \int_E |\rho'(x)|\, dx,$$

where $E = \Delta_k \cap \{x: |\rho'(x)| \geqq t\}$, $\|\rho'\|_{\Delta_k} = \max_{x\in\Delta_k} |\rho'(x)|$. Similarly

$$\int_{\Delta_k} \Phi(|\varphi'_{r+1,n}(x)|)\, dx = \int_0^{\|\varphi_{r+1,n}\|_{\Delta_k}} d\left\{\frac{\Phi(t)}{t}\right\} \int_e |\varphi'_{r+1,n}(x)|\, dx,$$

$$e = \Delta_k \cap \{x: |\varphi'_{r+1,n}(x)| \geqslant t\}.$$

Since $m\Delta_k \geqq \pi n^{-1}$, we have $\|\varphi'_{r+1,n}\|_{\Delta_k} = \|\varphi_{r,n}\|_{\Delta_k} = \|\varphi_{r,n}\|_C$, and by Favard's inequality $\|\rho'\|_{\Delta_k} \leqq \|\varphi'_{r+1,n}\|_{\Delta_k}$. Hence it is enough to establish

$$\int_E |\rho'(x)|\, dx \leqslant \int_e |\varphi'_{r+1,n}(x)|\, dx \tag{8}$$

for some t, $0 \leqq t \leqq K_{r,n}$. Here the integral on the left is the variation of $\rho(x)$ on the set of nonoverlapping intervals of Δ_k on which $|\rho'(x)| \geqq t$. Since $\rho(x)$ is monotonic [$\rho'(x) = R(x)$ has no zeros] on Δ_k, it is geometrically obvious that

$$\int_E |\rho'(x)|\, dx = \sum_{j=1}^{m} md_j,$$

where $\{md_j\}$ are the lengths of nonoverlapping intervals on the axis of ordinates. Since $m\Delta_k \geqq \pi n^{-1}$,

$$\int_e |\varphi'_{r+1,n}(x)|\, dx \geqslant \int_{\bar e} |\varphi'_{r+1,n}(x)|\, dx = 2\,|\varphi_{r+1,n}(\bar x)|.$$

By Kolmogorov's inequality $d_j \subseteq [-|\varphi_{r+1,n}(\bar x)|, |\varphi_{r+1,n}(\bar x)|]$. In fact, if we suppose the contrary, for example that $\beta_j \geqq |\varphi_{r+1,n}(\bar x)|$, $d_j = [\alpha_j, \beta_j]$, we draw the line $y = \beta_j$. Let the graphs of $|\rho(x)|$ and $|\varphi_{r+1,n}(x)|$ meet the line $y = \beta_j$ at ζ_1 and ζ_2: $\beta_j = |\rho(\zeta_1)| = |\varphi_{r+1,n}(\zeta_2)|$. Since $\|\rho^{(r+1)}\|_C = \|\varphi^{(r+1)}_{r+1,n}\|_C = 1$, Favard's inequality $\|\rho\|_C \leqq \|\varphi_{r+1,n}\|_C$ guarantees the existence of ζ_2. On the other hand, $|\rho'(\zeta_1)| = t > |\varphi'_{r+1,n}(\zeta_2)|$, which contradicts Kolmogorov's inequality for $\rho(x)$ and $\varphi_{r+1,n}(x)$ at the level $y = \beta_j$. Therefore

$$\sum_{j=1}^{m} md_j \leqslant 2\,|\varphi_{r+1,n}(\bar x)|.$$

We have proved (8), and consequently (7). Summing on k and combining the result with (5), we obtain (3), and Theorem 1 is established.

Theorem 2. *Let L^*_Φ be an Orlicz space with the Luxemburg norm. Then for all positive integers n and r we have*

$$\|f - S_{n-1}(f)\|_{(\Phi)} \leqslant n^{-r} \|\varphi_r\|_{(\Phi)} \|f^{(r)} - S_{n-1}(f^{(r)})\|_C,$$

where there is equality for $f(x) = \varphi_{r,n}(x)$.

Proof. We shall use the following inequalities (see for example [4], p. 97, Theorem 9.5): if $u(x) \in L^*_\Phi$ and $\|u\|_{(\Phi)} \leqq 1$, then

$$\int_0^{2\pi} \Phi(|u(x)|)\,dx \leqslant \|u\|_{(\Phi)}; \tag{9}$$

if $\|u\|_{(\Phi)} > 1$, then

$$\int_0^{2\pi} \Phi(|u(x)|)\,dx \geqslant \|u\|_{(\Phi)}. \tag{10}$$

For each $\lambda > 0$ the function $\Phi(\lambda t)$ is an N-function for $t \geqq 0$; hence by Theorem 1

$$\sup_{\|f^{(r)} - S_{n-1}(f^{(r)})\|_C \leqslant 1} \int_0^{2\pi} \Phi\{\lambda |f(x) - S_{n-1}(f, x)|\}\,dx = \int_0^{2\pi} \Phi\{\lambda |\varphi_{r,n}(x)|\}\,dx.$$

Take $\lambda = \lambda^* > 0$ so that $\|\lambda^* \varphi_{r,n}\|_{(\Phi)} = 1$. Then

$$\int_0^{2\pi} \Phi(\lambda^* |\varphi_{r,n}(x)|)\,dx = 1.$$

In fact, by (9),

$$\int_0^{2\pi} \Phi(\lambda^* |\varphi_{r,n}(x)|)\,dx \leqslant \|\lambda^* \varphi_{r,n}\|_{(\Phi)} = 1,$$

and if this were a strict inequality, then by the continuity of the integral with respect to λ there would be an $\epsilon > 0$ such that

$$\int_0^{2\pi} \Phi\{(\lambda^* + \varepsilon) |\varphi_{r,n}(x)|\}\,dx < 1. \tag{11}$$

However, $\|(\lambda^* + \epsilon)\varphi_{r,n}\|_{(\Phi)} > 1$ and by (10) we would have

$$\int_0^{2\pi} \Phi\{(\lambda^* + \varepsilon) |\varphi_{r,n}(x)|\}\,dx \geqslant \|(\lambda^* + \varepsilon)\varphi_{r,n}\|_{(\Phi)} > 1,$$

which contradicts (11). Hence we have established that

$$\sup_{\|f^{(r)}-S_{n-1}(f^{(r)})\|_C\leqslant 1} \int_0^{2\pi} \Phi\{\lambda^* | f(x) - S_{n-1}(f, x)|\} dx = 1. \tag{12}$$

We now show that

$$\sup_{\|f^{(r)}-S_{n-1}(f^{(r)})\|_C\leqslant 1} \|\lambda^*\{f - S_{n-1}(f)\}\|_{(\Phi)} = \|\lambda^*\varphi_{r,n}\|_{(\Phi)} = 1,$$

which, since $\lambda^* \neq 0$, is equivalent to the conclusion of Theorem 2. In fact, in the contrary case there would be a function $f_*(x)$ satisfying

$$\|f_*^{(s)} - S_{n-1}(f_*^{(r)})\|_C = 1, \quad \|\lambda^*\{f_* - S_{n-1}(f_*)\}\|_{(\Phi)} > 1.$$

Then by (10)

$$\int_0^{2\pi} \Phi\{\lambda^* | f_*(x) - S_{n-1}(f_* x) | dx > 1,$$

and we would have a contradiction with (12).

This completes the proof of Theorem 2.

Theorem 3. *Let L_Φ^* be an Orlicz space. Then for all positive integers n and r we have*

$$E_{n-1}(f, L) \leqslant n^{-r} \|\varphi_r\|_{(\Psi)} E_{n-1}(f^{(r)}, L_\Phi),$$

where $\Psi(t)$ is the N-function complementary to $\Phi(t)$, and the constant $n^{-r}\|\varphi_r\|_{(\Psi)}$ is best possible.

Proof. We shall need the equation (see for example [4], pp. 157 and 98)

$$\sup_{\|v\|_\Phi\leqslant 1} \int_0^{2\pi} u(x) v(x) dx = \|u\|_{(\Psi)},$$

which implies Hölder's inequality

$$\int_0^{2\pi} u(x) v(x) dx \leqslant \|u\|_{(\Psi)} \|v\|_\Phi.$$

In finding an upper bound for the best approximation

$$E_{n-1}(f, L) = \|f - T_{n-1}^*\|_L = \int_0^{2\pi} \{f(x) - T_{n-1}^*(x)\} \operatorname{sign}\{f(x) - T_{n-1}^*(x)\} dx,$$

we may restrict ourselves to functions $f(x)$ such that $\operatorname{sign}\{f(x) - T_{n-1}^*(x)\}$ is orthogonal to the first $2n-1$ harmonics, since all trigonometric polynomials have this property. If $F_r(x)$ is orthogonal to the first $2n-1$ harmonics, $F_r^{(r)}(x) = \operatorname{sign}\{f(x) - T_{n-1}(x)\}$, and $t_{n-1}^*(x)$ is the polynomial of best approximation to $f^{(r)}(x)$ in the L_Φ^* metric, then by integrating by parts r times and applying Hölder's inequality we obtain

$$E_{n-1}(f, L) = \int_0^{2\pi} \{f(x) - T^*_{n-1}(x)\} \operatorname{sign}\{f(x) - T^*_{n-1}(x)\}\, dx$$

$$= \int_0^{2\pi} \{f^{(r)}(x) + t^*_{n-1}(x)\} F_r(x)\, dx \leqslant E_{n-1}(f^{(r)}, L^*_\Phi)\|F_r\|_{(\Psi)}$$

$$\leqslant \|\varphi_{r,n}\|_{(\Psi)} E_{n-1}(f^{(r)}, L^*_\Phi).$$

The last inequality follows from Theorem 2. We now show that the constant $\|\varphi_{r,n}\|_{(\Psi)}$ is best possible.

For arbitrary $\epsilon > 0$ there is a function $v^*(x) = v^*(x, \epsilon)$ such that $\|v^*\|_\Phi \leqq 1$ and

$$\|\varphi_{r,n}\|_{(\Psi)} = \sup_{\|v\|_\Phi \leqslant 1} \int_0^{2\pi} \varphi_{r,n}(x)\, v(x)\, dx \leqslant \int_0^{2\pi} \varphi_{r,n}(x)\, v^*(x)\, dx + \varepsilon. \tag{13}$$

Let $a_0/2$ be the constant term of the Fourier series of $v^*(x)$, then $v_r^{*(r)}(x) = v^*(x) - a_0/2$; let $\tau^*_{n-1}(x)$ be the polynomial of best approximation to $v_r^*(x)$ in the L metric. Then, integrating by parts r times, we obtain

$$\int_0^{2\pi} \varphi_{r,n}(x)\, v^*(x)\, dx = \int_0^{2\pi} \operatorname{sign}\sin nx\, v_r^*(x)\, dx$$

$$= \int_0^{2\pi} \operatorname{sign}\sin nx \{v_r^*(x) - \tau^*_{n-1}(x)\}\, dx \leqslant E_{n-1}(v_r^*, L)$$

Thus for each $\epsilon > 0$ we have found a function $v^*(x) = v^*(x, \epsilon)$ satisfying $\|v^*\|_\Phi \leqq 1$ and

$$\|\varphi_{r,n}\|_{(\Psi)} \leqslant E_{n-1}(v_r^*, L) + \varepsilon.$$

This completes the proof of Theorem 3.

Corollary. *Under the hypotheses of Theorem 3 we have*

$$\sup_{\|f^{(r)}\|_\Phi \leqslant 1} E_{n-1}(f, L) = n^{-r}\|\varphi_r\|_{(\Psi)}.$$

Proof. From Theorem 3 we have at once

$$E_{n-1}(f, L) \leqslant n^{-r}\|\varphi_r\|_{(\Psi)} E_{n-1}(f^{(r)}, L^*_\Phi) \leqslant n^{-r}\|\varphi_r\|_{(\Psi)}\|f^{(r)}\|_\Phi,$$

and hence we only have to prove that the constant $n^{-r}\|\varphi_r\|_{(\Psi)}$ is best possible. Returning to (13), we show that $v^*(x)$ can be taken to be orthogonal to a constant. This fact lets us consider $v^*(x)$ as the rth derivative of some function: $f_\epsilon^{(r)}(x) = v^*(x)$. The rest of the proof is similar to the last part of the proof of Theorem 3.

First consider the case of even r. The graph of $\varphi_{r,n}(x)$ resembles the graph of $\pm\sin nx$. In fact, $\varphi_{r,n}(x)$ is an odd function of period $2\pi n^{-1}$. Consequently for each k we have

$$\int_0^{2\pi}\Psi\left\{\frac{|\varphi_{r,n}(x)|}{k}\right\}dx=\int_0^{2\pi}\Psi\left\{\frac{|\varphi_{r,n}[t(2n)^{-1}]|}{k}\right\}dt.$$

This means that $\varphi_{r,n}(x)$ and $\tilde{\varphi}_{r,n}(x)=\varphi_{r,n}[t(2n)^{-1}]$ have the same Luxemburg norm over a period:

$$\|\tilde{\varphi}_{r,n}\|_{(\Psi)}=\|\varphi_{r,n}\|_{(\Psi)}=\inf_{\rho(\varphi_{r,n},\Psi)\leqslant 1}k.$$

On the other hand, for each $\epsilon>0$ there is a function $\tilde{v}(x)=\tilde{v}(x,\epsilon)$ such that $\|\tilde{v}\|_{\Phi}\leqq 1$ and

$$\|\tilde{\varphi}_{r,n}\|_{(\Psi)}=\sup_{\|v\|_{\Phi}\leqslant 1}\int_0^{2\pi}\tilde{\varphi}_{r,n}(x)v(x)\,dx\leqslant\int_0^{2\pi}\tilde{\varphi}_{r,n}(x)\tilde{v}(x)\,dx+\varepsilon.$$

We may take $\operatorname{sign}\varphi_{r,n}(x)\,|\tilde{v}(2nx)|$ as $v^*(x)$, since this function is orthogonal to a constant and

$$\|\varphi_{r,n}\|_{(\Psi)}=\|\tilde{\varphi}_{r,n}\|_{(\Psi)}\leqslant\int_0^{2\pi}\tilde{\varphi}_{r,n}(x)\tilde{v}(x)\,dx+\varepsilon$$

$$=\int_0^{2\pi}\varphi_{r,n}(x)\operatorname{sign}\varphi_{r,n}(x)\,|\tilde{v}(2nx)|\,dx+\varepsilon.$$

If r is odd, we repeat the preceding discussion with the function $\varphi_{r,n}(x+\pi\{2n\}^{-1})$.

This completes the proof of the corollary.

We can establish the existence of functions for which there is equality in the inequality of Theorem 3. However, we shall not discuss this here.

References

[1] J. G. Favard, *Sur les meilleures procédés d'approximation de certaines classes de fonctions par des polynomes trigonométriques*, Bull. Sci. Math. **61** (1937), 209-224.

[2] Sun Yung-sheng, *On best approximation to classes of functions represented as convolutions*, Dokl. Akad. Nauk SSSR **118** (1958), 247-250. (Russian) MR **21** # 2144

[3] S. B. Stečkin, *On best approximation to conjugate functions by trigonometric polynomials*, Izv. Akad. Nauk SSSR Ser. Mat. **20** (1956), 197-206. (Russian) MR **17**, 1079.

[4] M. A. Krasnosel'skiĭ and Ja. B. Rutickiĭ, *Convex functions and Orlicz spaces. Problems of contemporary mathematics* GITTL, Moscow, 1958; Fizmatgiz, Moscow, 1959; English transl., Noordhoff, Groningen, 1961; Internat. Mono. on Advanced Math. Phys., Hindustan, Delhi (to appear). MR **21** # 5144; MR **23** # A4016.

[5] S. M. Nikol'skiĭ, *Approximation of functions in the mean by trigonometric polynomials*, Izv. Akad. Nauk SSSR Ser. Mat. **10** (1946), 207-256. (Russian) MR **8**, 149.

[6] A. N. Kolmogorov, *On inequalities between the upper bounds of the successive derivatives of an arbitrary function on an infinite interval*, Učen. Zap. Moskov. Gos. Univ. Matematika **30** (1939), 3-13; English transl., Amer. Math. Soc. Transl. (1) **2** (1962), 233-243. MR **1**, 298.

ON JACKSON'S INEQUALITY IN L_2

N. I. ČERNYH

Let $f(x)$ be a continuous function of period 2π, not identically constant, let $E_n(f)$ be its best uniform approximation by trigonometric polynomials of order n, and let $\omega(\delta; f)$ be its modulus of continuity in the uniform metric. N. P. Korneĭčuk [1] showed that

$$E_n(f) < 1\omega\left(\frac{\pi}{n+1}; f\right) \quad (n = 0, 1, \ldots) \tag{1}$$

and that for each $\epsilon > 0$ and $n = 0, 1, \cdots$ there is a continuous function $f_n(x, \epsilon)$ of period 2π such that

$$E_n(f_n(x, \varepsilon)) > \left(1 - \frac{1}{2(n+1)} - \varepsilon\right)\omega\left(\frac{\pi}{n+1}; f_n(x, \varepsilon)\right),$$

i.e. he proved that the constant multiplying ω in (1) cannot be replaced by a smaller one if we demand that the inequality holds simultaneously for all continuous functions and all $n = 0, 1, \cdots$.

Up to the present there has been no similar result for any space $L_p(0, 2\pi)$ $(1 \leqq p < \infty)$. We shall solve the problem for $p = 2$.

Let $L_2(0, 2\pi)$ be the space of measurable functions of period 2π, of summable square on $(0, 2\pi)$. Put

$$\|f(x)\|_{L_2} = \left\{\frac{1}{\pi}\int_0^{2\pi} |f(x)|^2 dx\right\}^{1/2},$$

$$\omega(\delta; f)_{L_2} = \sup_{|t| \leqslant \delta} \|f(x+t) - f(x)\|_{L_2},$$

$$E_n(f)_{L_2} = \|f(x) - S_n(x; f)\|_{L_2},$$

where $S_n(x; f)$ are the partial sums of order n of the Fourier series of $f(x)$. With this notation our main result can be stated as follows.

THEOREM. *Let* $f(x) \in L_2(0, 2\pi)$, $f(x) \not\equiv \text{const.}$ *For* $n = 0, 1, \cdots$ *we have*

$$E_n(f)_{L_2} < \frac{1}{\sqrt{2}}\omega\left(\frac{\pi}{n+1}; f\right)_{L_2} \tag{2}$$

and there is a function $f_n(x) \in L_2(0, 2\pi)$ *for which the ratio*

$$E_n(f_n)_{L_2} / \omega\left(\frac{\pi}{n+1}; f_n\right)_{L_2}$$

is arbitrarily close to $1/\sqrt{2}$.

First we prove the following lemma on the zeros of functions with non-negative Fourier coefficients.

Lemma. *Let $N \geqq 1$ be an integer, let $\rho_k (k \geqq N)$ be real numbers, and $0 < \sum \rho_k^2 < \infty$. Then the function*

$$F_N(t) = \sum_{k=N}^{\infty} \rho_k^2 \cos kt$$

takes both positive and negative values on $(0, \pi/N)$.

Proof. Put

$$\varphi_N(t) = \begin{cases} -\sin Nt & \text{for } 0 \leqslant t \leqslant \frac{\pi}{N}, \\ 0 & \text{for } \frac{\pi}{N} \leqslant t \leqslant \pi. \end{cases}$$

Expanding this function in a Fourier series of the form

$$\varphi_N(t) = A_0 + \sum_{k=1}^{\infty} A_k \cos kt,$$

we easily see that $A_N = 0$ and

$$A_k = \frac{4N}{\pi} \cos^2 \frac{k\pi}{2N} \frac{1}{k^2 - N^2} \geqslant 0 \quad (k > N).$$

From this we obtain, by Parseval's equation,

$$\frac{1}{\pi} \int_0^{\pi} \varphi_N(t) F_N(t)\, dt = \sum_{k=N}^{\infty} \rho_k^2 A_k \geqslant 0.$$

If we suppose that $F_N(t) \geqq 0$ for $0 < t < \pi/N$ then, by the definition of $\varphi_N(t)$ and the fact that $F_N(t) > 0$ in a neighborhood of $t = 0$, we obtain the inequality

$$\int_0^{\pi} \varphi_N(t) F_N(t)\, dt < 0,$$

which contradicts the preceding one. This establishes the lemma.

Proof of the theorem. Let $f(x) \in L_2(0, 2\pi)$ and let

$$\frac{a_0}{2} + \sum_1^{\infty} (a_k \cos kx + b_k \sin kx)$$

be its Fourier series. Put $a_k^2 + b_k^2 = \rho_k^2$. Then

$$E_n^2(f)_{L_2} = \sum_{k=n+1}^{\infty} \rho_k^2$$

and

$$\| f(x+t) - f(x) \|_{L_2}^2 = 2 \sum_{k=1}^{\infty} \rho_k^2 (1 - \cos kt).$$

Since $1 - \cos kt \geqq 0$, we obtain

$$\omega^2\left(\frac{\pi}{n+1}; f\right)_{L_2} = 2 \sup_{|t|<\frac{\pi}{n+1}} \sum_{k=1}^{\infty} \rho_k^2 (1 - \cos kt)$$

$$\geqslant 2 \sup_{|t|<\frac{\pi}{n+1}} \sum_{k=n+1}^{\infty} \rho_k^2 (1 - \cos kt) = 2E_n^2(f)_{L_2}$$

$$- 2 \inf_{|t|<\frac{\pi}{n+1}} \sum_{k=n+1}^{\infty} \rho_k^2 \cos kt.$$

By the lemma, if $E_n(f)_{L_2} \neq 0$, then

$$\inf_{|t|<\frac{\pi}{n+1}} \sum_{k=n+1}^{\infty} \rho_k^2 \cos kt < 0$$

and consequently

$$\omega^2\left(\frac{\pi}{n+1}, f\right)_{L_2} > 2E_n^2(f)_{L_2} \quad (n \geqslant 0).$$

But if $E_f(f) = 0$ and $f \not\equiv$ const. this inequality is trivial. This establishes (2).

To show that (2) is exact,[1] we put

$$F_n(t, \varepsilon) = 2A^2 \sum_{l=1}^{\infty} \left(\frac{\sin nl\delta}{nl}\right)^2 \cos nl\, t = \begin{cases} 1 - \frac{A^2}{2} \frac{\pi}{n} t & \text{for } 0 \leqslant t \leqslant 2\delta, \\ -\varepsilon^2 & \text{for } 2\delta \leqslant t \leqslant \frac{\pi}{n}, \end{cases}$$

where $\delta = \pi\epsilon^2/n(1+\epsilon^2)$, $A = (1+\epsilon^2)n/\epsilon\pi$, $\epsilon > 0$ is sufficiently small, and $n \geqq 1$. This function satisfies the following conditions: it is orthogonal to all harmonics up to order $n-1$ inclusive; its Fourier coefficients are nonnegative; and

$$F_n(0, \varepsilon) = 1, \quad F_n(t, \varepsilon) \geqslant -\varepsilon^2 \left(t \in \left(0, \frac{\pi}{n}\right)\right).$$

Then, putting

$$f_n(x, \varepsilon) = \sum_{l=1}^{\infty} (\alpha_l \cos nlx + \beta_l \sin nlx),$$

where α_l and β_l are arbitrary numbers satisfying $\alpha_l^2 + \beta_l^2 = 2A^2((\sin nl\delta)/nl)^2$, we evidently obtain

$$\omega^2\left(\frac{\pi}{n}; f_n(x, \varepsilon)\right)_{L_2} = 2E_{n-1}^2(f_n(x, \varepsilon))_{L_2} - 2 \inf_{|t|<\frac{\pi}{n}} F_n(t, \varepsilon) = 2E_{n-1}^2(f_n(x, \varepsilon))_{L_2} + 2\varepsilon^2,$$

[1] Another proof of the lower bound, for a more general case, is given by V. I. Berdyšev in a paper in the present volume.

or, equivalently,

$$\frac{\omega^2\left(\frac{\pi}{n}, f_n(x,\varepsilon)\right)_{L_2}}{E_{n-1}^2(f_n(x,\varepsilon))_{L_2}} = 2 + 2\varepsilon^2.$$

This completes the proof of the theorem.

Corollary. *If* $f(x) \in L_2(0, 2\pi)$, $f(x) \not\equiv \text{const.}$, *and* $f^{(r)}(x) \in L_2(0, 2\pi)$ $(r > 0)$, *then for all* $n > 0$ *the best approximation* $E_n(f)_{L_2}$ *satisfies*

$$E_n(f)_{L_2} < \frac{1}{\sqrt{2}} \frac{1}{(n+1)^r} \omega\left(\frac{\pi}{n+1}; f^{(r)}\right)_{L_2}. \tag{3}$$

Proof. As is well known,

$$\sup_{f^{(r)} \in L_2(0, 2\pi)} \frac{E_n^2(f)_{L_2}}{E_n^2(f^{(r)})_{L_2}} = \sup_{\rho_k} \frac{\sum_{n+1}^{\infty} \rho_k^2}{\sum_{n+1}^{\infty} k^{2r} \rho_k^2} = \frac{1}{(n+1)^{2r}}.$$

The corollary follows from this and the theorem.

References

[1] N. P. Korneĭčuk, *The exact constant in the theorem of D. Jackson on the best uniform approximation of continuous periodic functions,* Dokl. Akad. Nauk SSSR **145** (1962), 514-515 = Soviet Math. Dokl. **3** (1962), 1040-1041. MR **27** #521.

APPROXIMATION OF FUNCTIONS BY POLYNOMIALS WITH CONSTRAINTS

N. I. CERNYH

Introduction

The problem of approximating functions by polynomials whose coefficients are related by given linear dependencies was first investigated by V. A. Markov [1] in an important special case where $f(x) \equiv 0$ and the constraint on the polynomial $p_n(x)$ has the form $p_n^{(r)}(x_0) = 1$. Many authors have since studied this problem. The majority of them have given most attention to the problems of uniqueness and the characteristics of the best approximating polynomials or polynomials deviating least from zero. We mention the monograph [2], the review articles [3, 4], and the articles [5] and [6] where references to other authors may be found.

The present work contains an investigation of the behavior of the best approximations of functions by polynomials p_n of order n satisfying the given relationship $\Psi_n(p_n) = \rho_n$ when $n \to \infty$. The case will be considered where the homogeneous additive functional Ψ_n is successive with respect to n, that is, satisfies the condition

$$\Psi_{n+1}(p_n) = \Psi_n(p_n) = \Psi(p_n) \quad (n = 0, 1, \ldots).$$

In §1 we establish some general facts about approximations by polynomials with constraints in an arbitrary normed linear space. It turns out that the behavior of the best approximations of the element f by polynomials with constraints Ψ depends on the behavior of the best approximations of the element f without constraints and on the values of the functionals Ψ in the best approximating polynomials $p_n(f)$. The results of this section are closely related to the author's article [7].

In §2 we present the definition of certain concepts and the results we shall require regarding the best approximations of functions by algebraic polynomials $p_n(x)$ of degree n without additional constraints. In the cases where this has not been done earlier, we prove that these results are sharp.

In §3 we investigate the behavior of the values $\Psi(\hat{p}_n)$ of the functionals Ψ on the sequence of polynomials $\{\hat{p}_n(x)\}$ converging in the metric of the space $L_p[-1, 1]$ $(p \geqq 1)$ or $C_{[-1,1]}$ $(p = \infty)$. The main consideration here will be investigation of the functionals Ψ having the form

$$\Psi_1(p_n) = \int_{-a}^{a} p_n^{(r)}(x)\, dh(x) \quad (a > 1, r \geqslant 0) \tag{1}$$

or in the form

$$\Psi_2(p_n) = \int_{-a}^{a} p_n^{(r)}(x)\,(a^2 - x^2)^{\alpha}\,\varphi(x)\,dx \quad (\alpha > -1,\ a > 1,\ r \geqslant 0), \tag{2}$$

with certain restrictions on the functions $h(x)$ and $\varphi(x)$, permitting us to calculate for $n \to \infty$ the exact order of growth of the norms $\|\Psi_1\|_{n,p}$ and $\|\Psi_2\|_{n,p}$ of the functionals Ψ_1 and Ψ_2 in the subspaces of the polynomials of degree n with the metric of $L_p[-1,1]$ $(1 \leqq p \leqq \infty)$. These restrictions are such that the functional Ψ_1 is a generalization of the functional

$$\Psi(p_n) = p_n^{(r)}(a) \quad (a > 1),$$

and Ψ_2 is the integration functional

$$\Psi(p_n) = \int_{-a}^{a} p_n(x)\,dx \quad (a > 1).$$

Using the same devices as were used in the author's paper [8][1] for the norms $\|\Psi_1\|_{n,p}$ and $\|\Psi_2\|_{n,p}$ when $p = 1$, $p = 2$, and $p = \infty$, it would be possible to find asymptotic formulas. However, the scope of the present article does not permit us to do this.

In addition to the functionals Ψ_1 and Ψ_2 analogous problems are considered in this section for the functionals Ψ for which the norm $\|\Psi\|_{n,p}$ increases as the power of n. Among these functionals we have, for example, the functionals

$$\Psi(p_n) = p_n^{(r)}(x_0) \quad (n = 0, 1, \ldots),$$

where $x_0 \in [-1, 1]$, and $r > 0$.

It turns out that as $n \to \infty$ the sequence $\{\Psi(\hat{p}_n)\}$ increases significantly more slowly than the sequence of norms $\|\Psi\|_{n,p}$, and its growth depends on the rate of convergence of the sequence $\{\hat{p}_n(x)\}$ and also on the smoothness of the limit function. This phenomenon was first discovered by M. Zamansky [9] and S. B. Stečkin [10] for the differential operators of trigonometric polynomials $t_n(x)$.

In §4 we investigate the best approximations $E_n(f, \Psi, \rho)_p$ of the function $f(x)$ in the metric of the space $L_p[-1,1]$ $(1 \leqq p \leqq \infty)$ by polynomials $p_n(x)$ with the additional constraints $\Psi(p_n) = \rho_n$, where Ψ is one of the functionals investigated in §3.

In Theorem 4.1 we present an inequality for $E_n(f, \Psi, \rho)_p$ analogous to the familiar Jackson inequality [11]. The results contained in Theorems 4.2—4.7

---

[1] We take this opportunity to mention that the formula for the constant G_7 in Theorem 3 of §3 in [8] should read

$$G_7 = 2^{-\frac{1}{p}}\,\mu r!\,E_p^{(r)}\left(2\cos\frac{a}{2}\right)^{\frac{1}{2p}-\frac{r}{2}}.$$

may be considered analogs of the results of N. I. Ahiezer ([12], 1947 ed., p. 235) and Walsh and Russell ([13], p. 366) with respect to polynomial approximation of analytical functions which are bounded in a given domain, or, correspondingly, functions whose kth derivatives belong to the class $E_{p'}$ with respect to some region (see §2) and on the boundary of this region in the metric of $L_{p'}(1 \leqq p' \leqq \infty)$ satisfy a Lipschitz condition of order $\beta\,(0 < \beta < 1)$ or the Zygmund condition.

All the formulas obtained in §4 for the best approximations $E_n(f, \Psi, \rho)_p$, excluding formula (4.1) for $a = b$, either are asymptotic (for individual functions) or are upper bounds which are exact with respect to order in the corresponding classes of functions when $n \to \infty$. Some of the results of this section were presented without proof in [14].

§1. General theorems of approximation with constraints

Let X be an infinite normed linear space and let $\{x_e\}_0^\infty$ be a linearly dependent system of elements from X. We denote by $X_n = \{p_n\}\ (n = 0, 1, \cdots)$ the $(n+1)$-subspace of X spanned by the system $\{x_e\}_0^n$. The best approximation of the element $f \in X$ by the polynomials $p_n \in X_n$ of order n will be

$$E_n(f) = \inf_{p_n \in X_n} \| f - p_n \|,$$

and any polynomial $p_n(f) \in X_n$, satisfying the equation

$$\| f - p_n(f) \| = E_n(f),$$

will be called a best polynomial approximation to the element f.

Let Ψ be a nonzero homogeneous additive functional defined on the join of all X_n, and let its norm in the subspace X_n equal $\| \Psi \|_n$, where if $\| \Psi \|_{n0} = 0$ we shall consider that $\rho_{n0} = 0$.

The best approximation of the element $f \in X$ by polynomials p_n of order n with the constraint generated by the functional Ψ and the quantity ρ_n will be

$$E_n(f, \Psi, \rho_n) = \inf_{\substack{p_n \in X_n \\ \Psi(p_n) = \rho_n}} \| f - p_n \|,$$

and any polynomial $p_n(f, \Psi, \rho_n) \in X_n$, satisfying the equation

$$\| f - p_n(f, \Psi, \rho_n) \| = E_n(f, \Psi, \rho_n)$$

and the condition $\Psi(p_n(f, \Psi, \rho_n)) = \rho_n$, will be the polynomial of the best approximation of the element f corresponding to the constraint (Ψ, ρ_n).

In order to avoid further stipulations we shall conditionally state that $0/0 = 0$ until the beginning of §2.

THEOREM 1.1. *For any element $f \in X$ and any numbers ρ_n and ρ_n', one has*

$$|E_n(f,\Psi,\rho_n)-E_n(f,\Psi,\rho_n')|\leqslant\frac{|\rho_n-\rho_n|}{\|\Psi\|_n},\tag{1.1}$$

$$\left|E_n(f,\Psi,\rho_n)-\frac{|\rho_n'-\rho_n|}{\|\Psi\|_n}\right|\leqslant E_n(f,\Psi,\rho_n').\tag{1.2}$$

Proof. For any polynomial $p_n^*\in X_n$ we have

$$\begin{aligned}\inf_{\Psi(p_n)=\rho_n}\|p_n^*-p_n\|&=\inf_{\Psi(q_n)=\Psi(p_n^*)-\rho_n}\|q_n\|\\&=|\Psi(p_n^*)-\rho_n|\inf_{\Psi(p_n)=1}\|p_n\|=\frac{|\Psi(p_n^*)-\rho_n|}{\|\Psi\|_n}\end{aligned}\tag{1.3}$$

Therefore

$$\begin{aligned}E_n(f,\Psi,\rho_n)&\leqslant\inf_{\Psi(p_n)=\rho_n}\{\|f-p_n(f,\Psi,\rho_n')\|+\|p_n(f,\Psi,\rho_n')-p_n\|\}\\&=E_n(f,\Psi,\rho_n')+\inf_{\Psi(p_n)=\rho_n}\|p_n(f,\Psi,\rho_n')-p_n\|\\&=E_n(f,\Psi,\rho_n')+\frac{|\Psi(p_n(f,\Psi,\rho_n')-\rho_n|}{\|\Psi\|_n}\\&=E_n(f,\Psi,\rho_n')+\frac{|\rho_n'-\rho_n|}{\|\Psi\|_n}.\end{aligned}\tag{1.4}$$

Changing the roles of ρ_n' and ρ_n, we obtain

$$E_n(f,\Psi,\rho_n')\leqslant E_n(f,\Psi,\rho_n)+\frac{|\rho_n'-\rho_n|}{\|\Psi\|_n},$$

which together with inequality (1.4) is equivalent to (1.1).

Furthermore, by the definition of $p_n(f,\Psi,\rho_n')$ and equation (1.6), we have

$$\begin{aligned}\frac{|\rho_n'-\rho_n|}{\|\Psi\|_n}&=\inf_{\Psi(p_n)=\rho_n}\|p_n(f,\Psi,\rho_n')-p_n\|\\&\leqslant\inf_{\Psi(p_n)=\rho_n}\{\|p_n(f,\Psi,\rho_n')-f\|+\|f-p_n\|\}=E_n(f,\Psi,\rho_n')\\&+E_n(f,\Psi,\rho_n),\end{aligned}$$

which together with inequality (1.4) gives (1.2). The theorem is proved.

In many cases the behavior of the best approximation of $E_n(f)$ is known. Therefore, when investigating the best approximations $E_n(f,\Psi,\rho_n)$ the following corollaries may turn out to be useful. In them $p_n(f)\in X_n$ is the best polynomial for the element $f\in X$.

Corollary 1.1. *For any* $f\in X$

$$|E_n(f,\Psi,\rho_n)-E_n(f)|\leqslant\frac{|\Psi(p_n(f))-\rho_n|}{\|\Psi\|_n},\tag{1.5}$$

$$\left|E_n(f,\Psi,\rho_n)-\frac{|\Psi(p_n(f))-\rho_n|}{\|\Psi\|_n}\right|\leqslant E_n(f).\tag{1.6}$$

Let us assume that $\rho_n' = \Psi(\rho_n(f))$. Then, obviously, $E_n(f, \Psi, \rho_n') = E_n(f)$ and our statements follow directly from Theorem 1.1.

Inequalities (1.5) and (1.6) indicate in particular that when $n \to \infty$

$$E_n(f, \Psi, \rho_n) = E_n(f) + O\left(\frac{|\Psi(p_n(f)) - \rho_n|}{\|\Psi\|_n}\right) \tag{1.5'}$$

and

$$E_n(f, \Psi, \rho_n) = \frac{|\Psi(p_n(f)) - \rho_n|}{\|\Psi\|_n} + O(E_n(f)). \tag{1.6'}$$

If $E_n(f)$ and $|\Psi(p_n(f)) - \rho_n| \, \|\Psi\|_n^{-1}$ are of different order when $n \to \infty$ these formulas determine the asymptotic behavior of the best approximations $E_n(f, \Psi, \rho_n)$. In the general case we have

COROLLARY 1.2. *For any* $f \in X$

$$\begin{aligned} \frac{1}{3}\left\{E_n(f) + \frac{|\Psi(p_n(f)) - \rho_n|}{\|\Psi\|_n}\right\} &\leqslant E_n(f, \Psi, \rho_n) \\ &\leqslant E_n(f) + \frac{|\Psi(p_n(f)) - \rho_n|}{\|\Psi\|_n}. \end{aligned} \tag{1.7}$$

From inequality (1.6) we have

$$\frac{|\Psi(p_n(f)) - \rho_n|}{\|\Psi\|_n} \leqslant E_n(f, \Psi, \rho_n) + E_n(f),$$

and since $E_n(f, \Psi, \rho_n) \geqq E_n(f)$, we see that

$$E_n(f) + \frac{|\Psi(p_n(f)) - \rho_n|}{\|\Psi_n\|} \leqslant 3E_n(f, \Psi, \rho_n),$$

which is equivalent to the first inequality of (1.7). The second inequality is contained in Corollary 1.1.

The inequalities (1.7) in the special case where $X = C_{(0,2\pi)}$, X_{2n} is the subspace of trigonometric polynomials of order n, and $\rho_n = 0$, were mentioned earlier in equivalent form by this author [7].

From formula (1.2) we have a lower bound for $E_n(f, \Psi, \rho_n)$: *for any element* $f \in X$ *and any polynomial* $\hat{p}_n \in X_n$ *the inequality*

$$E_n(f, \Psi, \rho_n) \geqslant \frac{|\Psi(\hat{p}_n) - \rho_n|}{\|\Psi\|_n} - \|f - \hat{p}_n\| \tag{1.8}$$

is valid. To obtain this estimate it is sufficient to set $\rho_n' = \Psi(\hat{p}_n)$ in (1.2) and consider that

$$E_n(f, \Psi, \rho_n') \leqslant \|f - \hat{p}_n\|.$$

The inequality (1.8) is rougher than (1.2), but permits arbitrariness in the selection of the polynomial $\hat{p}_n$ which we shall use in §4.

The best approximating polynomial $p_n(f, \Psi, \rho_n)$ may be represented in the form

$$p_n(f, \Psi, \rho_n) = p_n(f) - [\Psi(p_n(f)) - \rho_n]\, q_n(f, \Psi),$$

where $q_n(f, \Psi) \in X_n$ and $\Psi(q_n(f, \Psi)) = 1$. Let us construct a similar polynomial

$$\overline{p}_n(f, \Psi, \rho_n) = p_n(f) - [\Psi(p_n(f)) - \rho_n]\, q_n^*(\Psi),$$

where $q_n^*(\Psi) \in X_n$ is the polynomial for which $\Psi(q_n^*) = 1$ and in which the norm $\|\Psi\|_n$ is obtained, that is, $\|q_n^*(\Psi)\| = \|\Psi\|_n^{-1}$. *The polynomial* $\overline{p}_n(f, \Psi, \rho_n)$ *satisfies the condition*

$$\Psi(\overline{p}_n(f, \Psi, \rho_n)) = \rho_n,$$

and its deviation

$$\Delta_n(f, \Psi, \rho_n) = \|f - \overline{p}_n(f, \Psi, \rho_n)\|$$

from the element $f \in X$ *when* $n \to \infty$ *is of the same order as the best approximation* $E_n(x, \Psi, \rho_n)$. *If* $E_n(f)$ *and* $|\Psi(p_n(f)) - \rho_n| \cdot \|\psi\|_n^{-1}$ *are quantities of different orders, then*

$$E_n(f, \Psi, \rho_n) \approx \Delta_n(f, \Psi, \rho_n),$$

where the symbol $\approx$ *denotes asymptotic equality.*

In fact, since

$$E_n(f, \Psi, \rho_n) \leqslant \Delta_n(f, \Psi, \rho_n)$$

and

$$\left| E_n(f) - \frac{|\Psi(p_n(f)) - \rho_n|}{\|\Psi\|_n} \right| \leqslant \Delta_n(f, \Psi, \rho_n) \leqslant E_n(f) + \frac{|\Psi(p_n(f)) - \rho_n|}{\|\Psi\|_n},$$

our assertion follows from Corollaries 1.1 and 1.2.

Let us also note that *if* $E_n(f, \Psi, \rho_n) \to 0$, $n \to \infty$, *the polynomials* $p_n(f, \Psi, \rho_n)$ *and* $\overline{p}_n(f, \Psi, \rho_n)$ *are asymptotically similar.*

Indeed, if $f = \theta$ (the zero element in X), it is easy to see that for all n

$$p_r(f, \Psi, \rho_n) = \overline{p}_n(f, \Psi, \rho_n) = \rho_n q_n^*(\Psi),$$

while if $f \neq \theta$, then $\|p_n(f, \Psi, \rho_n)\| \not\to 0$,

$$\overline{p}_n(f, \Psi, \rho_n) = p_n(f, \Psi, \rho_n) + \varepsilon_n.$$

and

$$\|\varepsilon_n\| \leqslant E_n(f, \Psi, \rho_n) + \Delta_n(f, \Psi, \rho_n) \leqslant 4E_n(f, \Psi, \rho_n) \to 0, \ n \to \infty.$$

Let us now assume that the set of polynomials p_n $(n = 0, 1, \cdots)$ is compact everywhere in the space X, i.e. that $\overline{\bigcup_n X_n} = X$, and let us see under what conditions imposed on the constraint (Ψ, ρ_n) we have $E_n(f, \Psi, \rho_n) \to 0$ for any element $f \in X$, when $n \to \infty$.

THEOREM 1.2. *Let* $\overline{\bigcup_n X_n} = X$. *Then in order that* $E_n(f, \Psi, \rho_n) \to 0$ $(n \to \infty)$ *for any element* $f \in X$, *it is necessary and sufficient that*

$$\|\Psi\|_n \to \infty,$$
$$\rho_n = o(\|\Psi\|_n) \quad (n \to \infty).$$

Before we prove this theorem let us note that *if the sequence of norms* $\|\Psi\|_n (n = 0, 1, \cdots)$ *is bounded, then* $E_n(f, \Psi, \rho_n)$ *can approach zero only in the case where* $\{\rho_n\}_{n=0}^{\infty}$ *is a converging sequence, and when this condition is fulfilled* $E_n(f, \Psi, \rho_n) \to 0$ *only for the elements f which satisfy the condition*

$$\Psi(f) = \lim_{n\to\infty} \rho_n.^2$$

Let us prove this. Let $\|\Psi\|_n \leqq M (n = 0, 1, \cdots)$ and let $E_n(f, \Psi, \rho_n) \to 0$ for some element $f \in X$. On the basis of the condition $\overline{\bigcup_n X_n} = X$, the sequence $\{E_n(f)\}$ approaches zero. From inequality (1.6) it then follows that $|\Psi(p_n(f)) - \rho_n| \to 0$. Since in addition

$$|\Psi(f) - \Psi(p_n(f))| \leqslant ME_n(f),$$

we see that the sequence $\{\rho_n\}$ converges and

$$\Psi(f) = \lim_{n\to\infty} \rho_n.$$

On the other hand, if $\rho_n \to \rho$ and $\Psi(f) = \rho$, then, selecting the fixed number m such that $\|\Psi\|_m \neq 0$, when $n \geqq m$, we obtain

$$\frac{|\Psi(p_n(f)) - \rho_n|}{\|\Psi\|_n} \leqslant \frac{|\Psi(p_n(f) - f) + \rho - \rho_n|}{\|\Psi\|_n} \leqslant \frac{ME_n(f) + |\rho - \rho_n|}{\|\Psi\|_m}.$$

From this and (1.5) we find that $E_n(f, \Psi, \rho_n) \to 0$ $(n \to \infty)$.

Let us now proceed to the proof of Theorem 1.2. Let $\|\Psi\|_n \to \infty$ and $\rho_n = o(\|\Psi\|_n)$. Let us then prove that for any $f \in X$ the sequence $\{E_n(f, \Psi, \rho_n)\}$ approaches zero. Let us first set $\rho_n = 0$ and assume that $E_n(f, \Psi, 0)$ does not approach zero for any element f. Then on the basis of the criterion (see for example [15], 1951 ed., pp. 164-165) for the density of a linear manifold in X we can find a nonzero bounded linear functional F which is defined in all of X, and such that $F(q_n) = 0$ any time that $\Psi(q_n) = 0$, $q_n \in X_n$. Since Ψ is a nonzero functional in the linear manifold $\bigcup_n X_n$, we have thus found the polynomial $\bar{p}_m$ for which $\Psi(\bar{p}_m) = 1$. Let us set $r_n = p_n - \Psi(p_n)\bar{p}_m$, where p_n is any polynomial from X_n. Obviously, $\Psi(r_n) = 0$. Consequently $F(r_n) = 0$, and thus $F(p_n) = \Psi(p_n)F(\bar{p}_m)$ for any polynomial $p_n \in X_n$ $(n = 0, 1, \cdots)$. Hence if $F(\bar{p}_m) \neq 0$ we have $F(p_n) = \lambda\Psi(p_n)$, $\lambda \neq 0$, which contradicts the boundedness of F. If $F(\bar{p}_m) = 0$, then $F(p_n) \equiv 0$, and on the basis of the premise that $\overline{\bigcup X_n} = X$ we have $F(f) = 0$ for any element $f \in X$, but this again contradicts the definition of F. The contradiction obtained proves that in the case of an unbounded sequence $\{\|\Psi\|_n\}$ for any element $f \in X$ we have $E_n(f, \Psi, 0) \to 0$.

Assuming in inequality (1.6) that $\rho_n = 0$, we find that for any $f \in X$

[2] In this case Ψ is a linear functional defined by continuity in the entire space X, and $\|\Psi\|_X \leqq M$.

$$\Psi(p_n(f)) = o(\|\Psi\|_n) \quad (n \to \infty), \tag{1.9}$$

if $\|\Psi\|_n \to \infty$. Then from inequality (1.5) it follows that $E_n(f, \Psi, \rho_n) \to 0$ for any elements $f \in X$, if $\|\Psi\|_n \to \infty$ and $\rho_n = o(\|\Psi\|_n)$. Q.E.D.

Let us prove the necessity of the conditions of the theorem. Let $E_n(f, \Psi, \rho_n) \to 0$ $(n \to \infty)$ for any $f \in X$. Assume that in this case $\|\Psi\|_n \leqq M < \infty$. Then, as has been proved above, for any $f \in X$

$$\Psi(f) = \lim_{n\to\infty} \rho_n,$$

which is possible only in the case where $\Psi(f) \equiv 0$; but this contradicts the fact that $\Psi(p_n)$ is a nonzero functional. Consequently the sequence $\{\|\Psi\|_n\}$ is unbounded. At the beginning of the proof of the theorem (see (1.9)) we showed that this implies $\Psi(p_n(f)) = o(\|\Psi\|_n)$. From this fact and from the inequality (1.6) we find that $\rho_n = o(\|\Psi\|_n)$. The theorem is proved.

From this theorem it follows in particular that in any converging sequence of polynomials $\hat{p}_n \in X_n$ $(n = 0, 1, \cdots)$ the functional Ψ cannot increase as rapidly as its norm.

COROLLARY 1.3. *Let* $\overline{\bigcup_n X_n} = X$ *and* $\|\Psi\|_n \to \infty$. *Then for any converging sequence of polynomials* $\hat{p}_n \in X_n$ $(n = 0, 1, \cdots)$ *one has*

$$\Psi(\hat{p}_n) = o(\|\Psi\|_n) \quad (n \to \infty). \tag{1.10}$$

If $\hat{p}_n = p_n(f)$ this equality coincides with (1.9), which has already been proved. In the general case the equality follows from the estimate (1.8) when $\rho_n = 0$ and the fact that under the conditions of the corollary $E_n(f, \Psi, 0) \to 0$ for any $f \in X$.

Let us present an interesting corollary of equality (1.10) which apparently has not been published before. *Let* $X = C_{[-1,1]}$, $\{x_k\} = \{x^k\}$ *and*

$$E_n(f) = \left\| f(x) - \sum_{k=0}^{n} c_k^{(n)}(f)\, x^k \right\|_{C_{[-1,1]}}.$$

Then, although the coefficient $c_m^{(n)}(f)$ *may approach infinity when* $n \to \infty$ *for any fixed* $m \geqq 1$ *and any function* $f(x) \in C_{[-1,1]}$, *the equality*

$$c_m^{(n)}(f) = o(n^m) \quad (n \to \infty) \tag{1.11}$$

remains valid nevertheless.

Let us note that directly from V. A. Markov's inequality [1]

$$|p_n^{(m)}(0)| \leqslant n^m \|p_n(x)\|_{C_{[-1,1]}} \quad (n > m)$$

it follows only that

$$c_m^{(n)}(f) = O(n^m).$$

If the derivative $f^{(r)}(x)$ exists for the function $f(x)$ and $f^{(r)}(x) \in \operatorname{Lip}\beta$ $(0 < \beta \leqq 1)$, then a more exact estimate will be given in place of (1.11) in §3.

Let us now mention the general scheme with the help of which it is frequently possible to obtain a more exact estimate than (1.10) for the sequence $\{\Psi(\hat{p}_n)\}$. In §3 we shall use this scheme in the space L_p for specific functionals. The inequalities (1.12) and (1.13) below were presented in the same general form by S. B. Stečkin and appear here with his permission.

Thus, let $\|\Psi\|_n \to \infty$, $f \in X$, $\hat{p}_n \in X_n$ and $\|f - \hat{p}_n\| \leqq F_n$ $(n = 0, 1, \cdots)$. Let us assume that $p_{n-1} = \theta$ and let $n_0 < n_1 < \cdots < n_s = n$. Then we have

$$|\Psi(\hat{p}_n)| = \left|\sum_{k=0}^{s} \Psi(\hat{p}_{n_k} - \hat{p}_{n_{k-1}})\right| \leqslant \sum_{k=0}^{s} \|\Psi\|_{n_k}(F_{n_{k-1}} + F_{n_k}), \qquad (1.12)$$

where $F_{n-1} = \|f\|$. If the sequence $\{\psi(p_n)\}$ converges and

$$A = \lim_{n\to\infty} \Psi(\hat{p}_n),$$

then analogously it is possible to estimate its rate of convergence: let us assume in this case $n = n_0 < n_1 < n_2 < \cdots$. Then

$$|\Psi(\hat{p}_n) - A| = \left|\sum_{k=0}^{\infty} \Psi(p_{n_{k+1}} - p_{n_k})\right| \leqslant \sum_{k=0}^{\infty} \|\Psi\|_{n_{k+1}}(F_{n_k} + F_{n_{k+1}}). \qquad (1.13)$$

If the sequence of indices $\{n_k\}$ is selected properly, good estimates may be obtained by this procedure. However, in some cases these estimates are quite rough. In §4, for example, a similar situation arises any time the function $f(x)$ has certain properties of smoothness which give the ordinal equality

$$E_n(f) \asymp \|\Psi\|_n^{-1}.$$

Knowing how to estimate the sequence $\{\Psi(p_n(f))\}$, with the help of Theorem 1.1 we can also obtain estimates of the approximations $E_n(f, \Psi, \rho_n)$.

In conclusion of this section let us note that along with $E_n(f, \Psi, \rho_n)$ it would be possible to consider the best approximations of the elements $f \in X$ by polynomials $p_n \in X_n$ for which $\Psi(p_n) \leqq \rho_n$ or $|\Psi(p_n)| \leqq \rho_n$, etc. However, these problems easily reduce to the preceding problem. For example,

$$\inf_{|\Psi(p_n)| \leqslant \rho_n} \|f - p_n\| = \inf_{|\lambda| \leqslant \rho_n} E_n(f, \Psi, \lambda).$$

Therefore we shall not consider them.

§2. Approximation of functions by algebraic polynomials without constraints

In this section we shall present certain facts from the theory of approximation of functions by algebraic polynomials, and also a few additional facts. The results and concepts presented here will be used in §§3 and 4 to investigate the behavior of approximating polynomials and the best approximations by polynomials with constraints.

Let $X = L_p[-1,1]$ $(1 \leqq p \leqq \infty)$ be the space of real functions integrable with the pth power in the interval $[-1,1]$ and

$$\| f(x) \|_{L_p[-1,1]} = \left\{ \frac{1}{2} \int_{-1}^{1} | f(x) |^p \, dx \right\}^{\frac{1}{p}};$$

when $p = \infty$ we shall consider that $X = L_\infty[-1,1] = C_{[-1,1]}$ is the space of real functions continuous in the interval $[-1,1]$ with a uniform metric; $X_n \subset X$ is the subspace of the algebraic polynomials $p_n(x)$ of degree n and correspondingly

$$E_n(f)_p = \inf_{p_n} \| f(x) - p_n(x) \|_{L_p[-1,1]} = \| f(x) - p_n(x; f, p) \|_{L_p[-1,1]},$$

$$E_n(f)_\infty = E_n(f), \quad p_n(x; f) = p_n(x; f, \infty).$$

Let us note that by the definition of the norm in the space $L_{p[-1,1]}$, for any p, p' $(1 \leqq p' \leqq p \leqq \infty)$ we have

$$E_n(f)_{p'} \leqslant E_n(f)_p \quad (f \in L_p[-1,1]).$$

We shall denote by $g(z)$ the function which is analytic in the plane with a slit in the interval $[-1,1]$, given by

$$g(z) = z + \sqrt{z^2 - 1};$$

we select the branch for which $\sqrt{z^2 - 1} > 0$ when $z > 1$. We shall denote by $T_n(z)$ the Čebyšev polynomials

$$T_n(z) = \cos n \arccos z = \frac{1}{2}(g^n(z) + g^{-n}(z)).$$

Throughout this article the letter C without indices will denote a constant. By a constant we mean a quantity which is independent of n. If we wish to emphasize that C may be selected dependent only on a given parameter λ, functional Ψ or class M, we shall write $C(\lambda)$, $C(\Psi)$, $C(M)$, $C(\lambda, \Psi)$, etc.

The accuracy with which it is possible to approximate the function $f(x)$ by polynomials of degree n will depend on the smoothness of $f(x)$. Thus, if $f(x)$ is continuously differentiable k times over the interval $[-1,1]$ and

$$\omega(\delta, f) = \sup_{\substack{|x' - x''| \leqslant \delta \\ x', x'' \in [-1,1]}} | f(x') - f(x'') |$$

is the continuity modulus of f in this interval, then, as Jackson proved [11],

$$E_n(f) \leqslant C(k) n^{-k} \omega\left(\frac{1}{n}; f^{(k)}\right) \quad (n > k). \tag{2.1}$$

This result may easily be carried over to the case of the best approximations in the mean: if for the function $f(x)$ the $(k-1)$th derivative is absolutely continuous in the interval $[-1,1]$; $f^{(k)}(x) \in L_p[-1,1]$ $(1 \leqq p \leqq \infty)$ and

$$\omega\left(\delta, f^{(k)}\right)_p = \sup_{|h| \leqslant \delta} \left\{ \int_{I_h} | f^{(k)}(x \mp h) - f^{(k)}(x) |^p \, dx \right\}^{\frac{1}{p}},$$

where $I_h = E(x : x, x+h) \in [-1, 1]$, then

$$E_n(f)_p \leqslant \frac{C(k, p)}{n^k} \omega\left(\frac{1}{n}; f^{(k)}\right)_p \quad (n > k). \tag{2.2}$$

The case of functions infinitely differentiable over the interval $[-1,1]$ was investigated by S. N. Bernšteĭn. He was the first to study the best approximations of functions which are analytic in the interval $[-1, 1]$. His basic results with respect to these questions may be found in the monograph **[16]**. The most complete result for analytic functions belongs to N. I. Ahiezer **[12]**. We shall present this result.

Let us denote by $G(b)$ the interior of the ellipse $\Gamma(b)$ with foci at the points $z = \pm 1$ passing through the point $z = b > 1$. If the function $f(x)$ is analytic in the interval $[-1, 1]$, it will also be analytic in some domain $G(b)$. If we denote by $MH(0,0,\infty, G(b))$ the class of functions $f(z)$ which are analytic in the domain $G(b)$ and real in the interval $[-1, 1]$, for which

$$|f(z)| \leqslant M \quad (z \in G(b)),$$

then Ahiezer's result may be formulated as follows:

$$\sup_{f \in MH(0,\,0,\,\infty,\,G(b))} E_{n-1}(f) = \frac{8M}{\pi} \sum_{l=0}^{\infty} \frac{(-1)^l}{(2l+1)} \frac{g^{(2l+1)n}(b)}{g^{2(2l+1)n}(b) + 1}. \tag{2.3}$$

Let $\{A_n\}$ and $\{B_n\}$ be two sequences which depend on certain parameters. Let us stipulate that

$$A_n \asymp B_n \quad (n \to \infty),$$

if there exist absolute constants $c_2 \geqq c_1 > 0$ such that

$$c_1 A_n \leqslant B_n \leqslant c_2 A_n$$

for all admissible values of the parameters. From formula (2.3), in particular, it follows that

$$\sup_{f \in MH(0,\,0,\,\infty,\,G(b))} E_n(f) \asymp \frac{C(M, b)}{g^n(b)}. \tag{2.4}$$

Walsh and Russell **[13]** investigated the rate at which the best approximations of analytic functions by polynomials $p_n(z)$ vanish. They further subdivided the class of functions which are analytic in the domain G according to their smoothness on the boundary of the domain. We shall first present some exact definitions of the classes of functions.

Let G be a domain in the complex plane bounded by the rectifiable Jordan curve γ. We say (see for example **[17]**, p. 203) that the function $f(z)$ in the domain G *belongs to the class* E_p $(p > 0)$ if it is analytic in G and if in G there is a sequence of curves $\{\gamma_n\}$ converging to γ (see [17], pp. 44-45) and such that

$$\left\{\int_{\gamma_n} |f(z)|^p |dz|\right\}^{1/p} \leqslant C.$$

Then let γ be an analytic Jordan curve; let s be the arc length on γ reckoned from some fixed point; let $p \geqq 1$; let $k \geqq 0$ be an integer; let $f^{(k)}_{(z)} \in E_p$ in the domain G and let $F(s) = f^{(k)}(\zeta)$ be the angular boundary values of the function $f^{(k)}(z)$ on the curve γ. We shall then say that the function $f(z)$ belongs to the class $MH(k,\beta,p,G)$ $(0 < \beta \leqq 1)$ if for any $h > 0$ the inequality

$$\left\{\int_{\gamma} |F(s+h) - F(s)|^p ds\right\}^{1/p} \leqslant Mh^{\beta} \quad (0 < \beta < 1),$$

is fulfilled; in the case $\beta = 1$ this is replaced by the inequality

$$\left\{\int_{\gamma} |F(s+h) - 2F(s) + F(s-h)|^p ds\right\}^{1/p} \leqslant Mh.$$

Analogously the condition $f(z) \in MH(k,\beta,\infty,G)$ indicates that the function $f(z)$ is analytic in the domain G and continuous in the closed domain $\overline{G}$, and the function $F(s) = f(z)|_{z\in\gamma}$ is continuously differentiable k times along γ, and that, for any $h > 0$,

$$|F^k(s+h) - F^{(k)}(s)| \leqslant Mh^{\beta} \qquad (0 < \beta < 1)$$

or correspondingly

$$|F^{(k)}(s+h) - 2F^{(k)}(s) + F^{(k)}(s-h)| \leqslant Mh \qquad (\beta = 1).$$

Later we shall consider approximation of the functions of the class $MH(k,\beta,G(b))$ by polynomials $p_n(x)$ in the interval $[-1,1]$, assuming in addition that the functions from this class are real in the interval $[-1,1]$. Let us note directly that all the results of this and subsequent sections with respect to analytic functions, except for equality (2.3), also hold with appropriate definition of the function $g(z)$ in the more general case where approximations are investigated on an arbitrary analytic Jordan curve γ_1, and the classes of functions (without additional restrictions of the type of reality) are defined by the domain G bounded by the level line corresponding to the curve γ_1, that is, the prototype of the curve $|w| = R > 1$ with conformal mapping of the exterior of γ_1 on the exterior of the unit curve $|w| = 1$ $(w(\infty) = \infty)$. The case which we have investigated is the one of greatest interest to us and with some simplification of the arguments allows us to see how one must formulate and prove the corresponding results in the general case.

As applied to the class $MH(k,\beta,p,G(b))$ the results of Walsh and Russell referred to above may be formulated in the following way. Let $M > 0, 0 < \beta \leqq 1$, $1 \leqq p \leqq \infty$, $b > 1$, and let $k \geqq 0$ be an integer; then there exists a constant $C(M) = C(M, k, \beta, p, b)$ such that for any function $f \in MH(k, \beta, p, G(b))$ we have

$$E_n(f) \leqslant C(M)\, n^{-k-\beta} g^{-n}(b) \quad (n \geqslant k). \tag{2.5}$$

Just as in the case of the class $MH(0, 0, \infty, G(b))$, it is impossible to improve this estimate in the sense of order when $n \to \infty$. Although Walsh and Russell themselves do not mention this, it is easy to establish by considering the following function, of Weierstrass type, as an example:

$$f_{k,\beta}(z) = \mu \sum_{l=1}^{\infty} \nu^{-l(k+\beta)} g^{-\nu^l}(b)\, T_{\nu^l}(z),$$

where $\nu > 1$ is an even number. However, it is also possible to prove a stronger statement.

THEOREM 2.1. *Let $k \geqq 0$ be a natural number, $M > 0$, $b > 1$, $0 < \beta \leqq 1$, $1 \leqq p \leqq \infty$. Then for any p' $(1 \leqq p' \leqq \infty)$ the following formula is correct:*

$$\sup_{f \in MH(k, \beta, p, G(b))} E_n(f)_{p'} \asymp \frac{C(M)}{n^{k+\beta} g^n(b)}. \tag{2.6}$$

PROOF. Since

$$E_n(f)_1 \leqslant E_n(f)_{p'} \leqslant E_n(f) \qquad (1 \leqslant p' \leqslant \infty)$$

on the basis of inequality (2.5) we have

$$\sup_{f \in MH(k, \beta, p, G(b))} E_n(f)_{p'} \leqslant C(M)\, n^{-k-\beta} g^{-n}(b),$$

and to prove formula (2.6) it is sufficient for any $n > 2$ to construct a function $f_n \in MH(k, \beta, p, G(b))$, for which

$$E_{n-2}(f_n)_1 \geqslant C n^{-k-\beta} g^{-n}(b) \quad (C > 0).$$

Let us assume

$$f_n(z) = \mu \sum_{l=0}^{\infty} n^{-l(k+\beta+1)} g^{-n^l}(b)\, T'_{n^l}(z) \quad (\mu > 0,\ n = 2, 3, \dots). \tag{2.7}$$

Let us show that $f_n(z) \in CH(k, \beta, \infty, G(b))$ for all $n > 2$, where $C = \mu C(k, \beta, b)$. From this it will follow that by choosing $\mu = \mu(p, k, \beta, b)$ properly we can ensure that $f_n(z) \in MH(k, \beta, p, G(b))$ $(1 \leqq p \leqq \infty)$ for all $n > 2$.

Obviously $f_n(z)$ is k times continuously differentiable in the closed region $\overline{G(b)}$. Since

$$|g(z)|_{z \in \Gamma_b} = g(b),$$

when $0<\beta<1$ for any points $z_1, z_2 \in \Gamma_b$, we obtain

$$\begin{aligned} &|f_n^{(k)}(z_1)-f_n^{(k)}(z_2)| \\ &\quad \leqslant 4\mu|z_1-z_2|\sum_{l=0}^{m} n^{-l(k+\beta+1)}g^{-n^l}(b)\max_{z\in\Gamma_b}|T_{n^l}^{(k+2)}(z)| \\ &\quad + 2\mu\sum_{l=m+1}^{\infty} n^{-l(k+\beta+1)}g^{-n^l}(b)\max_{z\in\Gamma_b}|T_{n^l}^{(k+1)}(z)| \\ &\quad \leqslant \mu C(k,b)\Big\{|z_1-z_2|\sum_{l=0}^{m} n^{l(1-\beta)}+\sum_{l=m-1}^{\infty} n^{-l\beta}\Big\} \\ &\quad \leqslant \mu C(k,b)\Big\{|z_1-z_2|\frac{n^{1-\beta}}{n^{1-\beta}-1}n^{m(1-\beta)}+\frac{1}{n^{(m+1)\beta}}\frac{n^{\beta}}{n^{\beta}-1}\Big\}. \end{aligned}$$

If $|z_1-z_2|\leqq 1$, then, selecting m such that

$$\frac{1}{n^{m+1}}\leqslant|z_1-z_2|\leqslant\frac{1}{n^m},$$

we obtain

$$|f_n^{(k)}(z_1)-f_n^{(k)}(z_2)|\leqslant\mu C(k,\beta,b)|z_1-z_2|^{\beta}$$

Obviously this inequality will be fulfilled in the case where $|z_1-z_2|>1$, $z_1, z_2\in\Gamma_b$, if the constant $C(k,\beta,b)$ is increased somewhat. Consequently, $f_n(z)\in\mu C(k,\beta,b)H(k,\beta,\infty,G(b))$ $(0<\beta<1)$. It is proved analogously that $f_n(z)\in\mu C(k,b)H(k,1,\infty,G(b))$ when $\beta=1$.

Let us now estimate the lower bound of $E_{n-2}(f)_1$. Since the systems $\{l^{-1}T_l'(x)\}_1^{\infty}$ and $\{\sin l\arccos x\}_1^{\infty}$ are biorthogonal in the interval $[-1,1]$, we obtain

$$\begin{aligned} \mu n^{-(k+\beta)}g^{-n}(b) &= \frac{2}{\pi}\int_{-1}^{1} f_n(x)\sin n\arccos x\,dx \\ &= \frac{2}{\pi}\int_{-1}^{1}\Big[f_n(x)-\sum_{l=1}^{n-1}c_lT_l'(x)\Big]\sin n\arccos x\,dx, \end{aligned}$$

and hence

$$\frac{\mu\pi}{4}n^{-(k+\beta)}g^{-n}(b)\leqslant\inf_{c_l}\frac{1}{2}\int_{-1}^{1}\Big|f_n(x)-\sum_{l=1}^{n-1}c_lT_l'(x)\Big|dx=E_{n-2}(f)_1.$$

Theorem 2.1 is proved.

Remark. In order to prove the sharpness of the estimate in the general formulation of the Walsh and Russell theorem (see [13], p. 366), it is necessary to consider in lieu of the sequence (2.7) the sequence of functions each of which is constructed analogously from the Faber polynomials (see for example [18]) corresponding to the region with respect to which the classes of functions are investigated.

We shall not use the Walsh and Russell results with respect to the classes of analytic functions $H(k,\beta,p)$ and $Z(k,p)$ introduced by them [13] for negative integral values of k. We shall use only the following classes of functions: by $MH(0,0,p,G(b))$ ($M>0$, $1 \leqq p < \infty$, $b>1$) we shall denote the set of functions from the class E_p with respect to the domain $G(b)$ for which

$$\left\{\int_{\Gamma_b} |f(\lambda)|^p |d\lambda|\right\}^{1/p} \leqslant M.$$

Formula (2.4) of N. I. Ahiezer may be extended to these classes.

THEOREM 2.2. *Let $M>0$, $b>1$, $1 \leqq p \leqq \infty$. Then for any p' ($1 \leqq p' \leqq \infty$)*

$$\sup_{f \in MH(0,\,0,\,p,\,G(b))} E_n(f)_{p'} \asymp C(M,p,b)\, g^{-n}(b). \tag{2.8}$$

PROOF. In the case $p=\infty$ the upper bound is obtained from formula (2.4). In the remaining cases it may be obtained by applying the Walsh-Russell scheme. This proof is quite short, so we will give it here.

Let $f \in MH(0,0,p,G(b))$. According to Fatou's Theorem $f(z)$ has integrable angular values $f(\lambda)$ almost everywhere in Γ_b. Let $p_n(z)$ be the interpolation polynomial of the function $f(z)$ with points of interpolation at the zeros of the Čebyšev polynomial $\cos(n+1)\arccos z$. By the subtraction theorem

$$f(z) - p_n(z) = \frac{1}{2\pi i}\int_{\Gamma_b} \frac{\cos(n+1)\arccos z}{\cos(n+1)\arccos\lambda}\,\frac{f(\lambda)}{\lambda - z}\,d\lambda \quad (z \in G(b)).$$

Applying Hölder's inequality, we obtain

$$\begin{aligned} E_n(f)_{p'} &\leqslant E_n(f) \leqslant \|f(x) - p_n(x)\|_{C[-1,1]} \\ &\leqslant \frac{1}{2\pi(b-1)}\left\{\int_{\Gamma_b} |f(\lambda)|^p |d\lambda|\right\}^{1/p} \left\{\int_{\Gamma_b} \frac{|d\lambda|}{|g^n(\lambda)|^q}\right\}^{1/q} \leqslant C(M)\, g^{-n}(b). \end{aligned}$$

To obtain the lower bound we may consider, for example, the sequence of functions

$$f_n(z) = \mu \sum_{l=0}^{\infty} n^{-l} g^{-nl}(b)\,(l+1)^{-2}\, T'_{nl}(z) \quad (\mu > 0,\ n = 2, 3, \ldots).$$

Since for $z \in \Gamma_b$ we have

$$|f_n(z)| \leqslant 2\mu\,(b^2-1)^{-\frac{1}{2}} \sum_{l=0}^{\infty} (l+1)^{-2},$$

for any p ($1 \geqq p \geqq \infty$), it is possible to select $\mu = \mu(p,b,M)$ such that all functions $f_n(z)$ ($n = 2,3,\cdots$) will belong to the class $MH(0,0,p,G(b))$. On the other hand, just as in the proof of Theorem 2.1, we obtain

$$C\mu g^{-n}(b) = \frac{1}{2}\int_{-1}^{1} f_n(x) \sin n \arccos x\, dx \leqslant E_{n-2}(f)_1 \leqslant E_{n-2}(f)_{p'}.$$

The theorem is proved.

Later we shall require the following two theorems for the classes $MH(k,\beta,p,G)$.

THEOREM 2.3 (WALSH-RUSSELL [13]). *Let G be the interior of the closed analytic curve Γ, and let $M>0$, $1\leqq p\leqq\infty$, $k\geqq 0$, $0<\beta\leqq 1$. There exists a number $C(M)=C(M,k,\beta,p,G)$, such that for any function $f(z)\in MH(k,\beta,p,G)$, and any natural number $n\geqq k$ one can find a polynomial $\pi_n(z)$ of degree n satisfying the inequality*

$$\|f(z)-\pi_n(z)\|_{L_p(\Gamma)}\leqslant C(M)\,n^{-k-\beta} \tag{2.9}$$

(when $p=\infty$ we shall take

$$\|f\|_{L_\infty(\Gamma)}=\|f\|_{C(\Gamma)}=\max_{z\in G}|f(z)|).$$

THEOREM 2.4 (HARDY-LITTLEWOOD, ZYGMUND). *Let G be the domain bounded by the closed analytic curve Γ; let $M>0$, $1\leqq p\leqq\infty$, $k\geqq 0$ an integer, $0\leqq\beta\leqq 1$ and $f(z)\in MH(k,\beta,p,G)$.*[3] *Denote by $\gamma\subset G$ the level line of the domain G for its conformal mapping on the unit circle $|w|\leqq 1$. If $s\geqq k+2$, then on contraction of the curve γ to the boundary Γ of G one obtains*

$$\|f^{(s)}\|_{L_p(\gamma)}=O(\rho^{\beta-s+k}(\gamma,\Gamma)), \tag{2.10}$$

$$\|f^{(s)}\|_{C(\gamma)}=O(\rho^{\beta-s+k-\frac{1}{p}}(\gamma,\Gamma)), \tag{2.11}$$

where $\rho(\gamma,\Gamma)$ is the distance between γ and Γ, and the constant entering into O does not depend on the function $f(z)$.

In the case where G is a unit circle the theorem was proved for $0\leqq\beta<1$ by Hardy and Littlewood, and for $\beta=1$ by Zygmund (see the bibliography in [13]).

The validity of (2.10) and (2.11) also follows easily in the general case, since with a conformal mapping of $z=z(w)$ of the domain G on the unit circle the function $f(z)\in MH(0,\beta,p,G)$ becomes $f(z(w))\in M_1H(0,\beta,p,|w|\leqq 1)$ (see [13], p. 358), where $M_1=M_1(M,\beta,p,G)$. It is true that it does not follow from the proof given by Walsh and Russell that M_1 is independent of the function $f(z)$, but their proof may easily be modified by replacing $f(z)$ by $f(z)-A(f)$ where the constant $A(f)$ is defined by the equality

$$\|f(z)-A(f)\|_{L_p(\Gamma)}=\inf_A\|f(z)-A\|_{L_p(\Gamma)}.$$

§3. The behavior of functionals of approximating polynomials

Let Ψ be a homogeneous additive functional defined in the linear manifold of all algebraic polynomials, and let $\|\Psi\|_{n,p}$ be its norm in the subspace of the polynomials $p_n(x)$ of degree n, taken in the metric of the space $L_p(-1,1)$ $(1\leqq p\leqq\infty)$; we set $\|\Psi\|_n=\|\Psi\|_{n,\infty}$.

---

[3] When $\beta=0$ this means that $\|f^{(k)}(z)\|_{L_p(\Gamma)}\leqq M$.

DEFINITION 3.1. Let $a > 1$ and let $r \geqq 0$ be an integer. We shall denote the functional Ψ by $\Psi_{a,r}$ if it can be represented in the form

$$\Psi(p_n) = \int_{-a}^{a} p_p^{(r)}(x)\, dh(x) = \Psi_{a,r}(p_n) \quad (n = 0, 1, \ldots), \tag{3.1}$$

where the function $h(x)$ does not depend on n, has bounded variation in the interval $[-a, a]$ and is not identically constant simultaneously in every interval $[-a, -a+\epsilon)$ $(a-\epsilon, a]$ for any $\epsilon \in (0, a)$.

The latter restriction does not restrict the class of functionals under investigation. It is imposed so that the functional $\Psi_{a,r}$ will be defined uniquely. If we say that a property is realized for any functional $\Psi_{a,r}$ this means that a and r are fixed numbers, and any function $h(x)$ may be selected as long as it satisfies the restrictions in the definition of $\Psi_{a,r}$. This also applies to the functionals $\Psi_a^{(r)}$ and $\Psi_{\alpha,a}^{(r)}$ defined below.

DEFINITION 3.2. We shall denote the functional $\Psi_{a,r}$ by $\Psi_a^{(r)}$ if its defining function $h(x)$ is constant for some $\delta > 0$ in the intervals $(-a, -a+\delta)$, $(a-\delta, a)$ and

$$[h(a) - h(a-o)]^2 + [h(-a) - h(a+o)]^2 \neq 0. \tag{3.2}$$

Let us note that all the results of this and the following section referring to the functionals $\Psi_a^{(r)}$ remain valid if the conditions on the function $h(x)$ are relaxed somewhat. For example, it is sufficient to require that $h(x)$ satisfy the condition (3.2) and be monotonic in the neighborhood of the points $x = \pm a$. Corollary 3.1 below is easily carried over to this case. However, the proofs of Theorems 3.2—3.5 and 4.3—4.5 are significantly more complicated.

DEFINITION 3.3. Let $a > 1$, $\alpha > -1$, and let $r \geqq 0$ be an integer. We shall denote the functional Ψ by $\Psi_{\alpha,a}^{(r)}$ if it may be represented in the form

$$\Psi(p_n) = \int_{-a}^{a} p_n^{(r)}(x)(a^2 - x^2)^\alpha \varphi(x)\, dx = \Psi_{\alpha,a}^{(r)}(p_n) \quad (n = 0, 1, \ldots),$$

where $\varphi(x)$ is integrable in the interval $[-a, a]$ and continuous at the points $x = \pm a$ and such that

$$\varphi^2(a) + \varphi^2(-a) \neq 0.$$

The functional $\Psi_{\alpha,a}^{(r)}$ is a special form of the functional $\Psi_{a,r}$ where

$$dh(x) = (a^2 - x^2)^\alpha \varphi(x)\, dx.$$

In this section we shall calculate the order of growth of the norms of the functionals $\Psi_{a,r}$, and study the behavior of these functionals in the sequences of polynomials $\{p_n(f)\}$ under the condition that $f(x)$ belongs to the given class and that the sequence of polynomials converges to $f(x)$ in the metric of $L_{p'}[-1,1]$ $(1 \leqq p' \leqq \infty)$ at a definite rate. The latter problem is also considered for arbitrary functionals Ψ satisfying the conditions

$$\| f^{\nu} \|_{n, p'} \asymp n^{\nu} \quad (\nu > 0)$$

Lemma 3.1. *Let $r \geqq 0$ be an integer, $a > 1$. For any functional $\Psi_{a,r}$ there is a number $\mu > 0$ such that for all $n = 1, 2, \cdots$ and all p $(1 \leqq p \leqq \infty)$*

$$\| \Psi_{a, r} \|_{n, p} \leqslant \mu n^r g^n (a). \tag{3.3}$$

This estimate in the class of all functionals $\Psi_{a,r}$ is exact in the sense of order when $n \to \infty$. As will be shown, it is attained for the functionals $\Psi_a^{(r)}$. However, for individual functionals it may turn out to be rough. Thus, for example (see Lemma 3.2),

$$\| \Psi^{(r)}_{\alpha, a} \|_{n,p} \asymp n^{r-\alpha-1} g^n (a) \quad (n \to \infty)$$

Proof. Let $1 < a_0 < a$. Integrating with the weight $|dh(x)|$ the inequality

$$| p_n^{(r)} (x) | \leqslant C (r, x) n^r g^n (x) \| p_n (\xi) \|_{L[-1, 1]} (| x | > 1), \tag{3.4}$$

([8], §2, Corollary 5), where

$$C (r, x) = C (r) \frac{g^3 (x)}{g^2 (x) - 1} (x^2 - 1)^{-\frac{r+1}{2}},$$

we obtain

$$\left| \left(\int_{-a}^{-a_0} + \int_{a_0}^{a} \right) p_n^{(r)} (x) \, dh (x) \right| \leqslant C n^r g^n (a) \| p_n (\xi) \|_{L[-1, 1]} \int_{-a}^{a} | dh (x) |.$$

Then, applying A. A. Markov's inequality [19]

$$\| p_n' (x) \|_{C[-a_0, a_0]} \leqslant a_0^{-1} n^2 \| p_n (x) \|_{C[-a_0, a_0]},$$

P. L. Čebyšev's [20]

$$| p_n (x) | \leqslant | T_n (x) | \| p_n (\xi) \|_{C[-1,1]} (| x | > 1) \tag{3.5}$$

and D. Jackson's (see for example [21])

$$\| p_n (x) \|_{C[-1,1]} \leqslant C n^2 \| p_n (x) \|_{L[-1,1]},$$

we obtain

$$\left| \int_{-a_0}^{a_0} p_n^{(r)} (x) \, dh (x) \right| \leqslant C \left\{ n^{2r+2} \int_{-1}^{1} | dh (x) | \right.$$
$$\left. + \, n^{2r+2} g^n (a_0) \left(\int_{-a_0}^{-1} | dh (x) | + \int_{1}^{a_0} | dh (x) | \right) \right\} \| p_n (\xi) \|_{L[-1,1]}$$
$$\leqslant C n^{2r+2} g^n (a_0) \| p_n (\xi) \|_{L[-1,1]}.$$

Since $g(a_0) < g(a)$, from this we find that

$$\left| \int_{-a}^{a} p_n^{(r)} (x) \, dh (x) \right| \leqslant C (\Psi_{a, r}) n^r g^n (a) \| p_n (\xi) \|_{L[-1,1]},$$

that is,

$$\|\Psi_{a,r}\|_{n,1} \leqslant C(\Psi_{a,r})\, n^r g^n(a).$$

From the definition of the norms in the space $L_p[-1,1]$ it follows that

$$\|\Psi_{a,r}\|_n \leqslant \|\Psi_{a,r}\|_{n,p} \leqslant \|\Psi_{a,r}\|_{n,1} \quad (1 < p < \infty).$$

The lemma is proved.

COROLLARY 3.1. *Let $r \geqq 0$ be an integer, and take $a > 1$. For any functional $\Psi_a^{(r)}$ there are numbers $0 < \mu_1 \leqq \mu < \infty$ such that for all p $(1 \leqq p \leqq \infty)$ and all $n = 1, 2, \cdots$*

$$\mu_1 n^r g^n(a) \leqslant \|\Psi_a^{(r)}\|_{n,p} \leqslant \mu n^r g^n(a) \tag{3.6}$$

The second of the inequalities (3.6) coincides with the estimate (3.3). The first inequality is obtained in the following way. On the basis of the conditions imposed on the function $h(x)$ we have

$$\begin{aligned}\Psi_a^{(r)}(T_m) &= \int_{-a}^{a} T_m^{(r)}(x)\, dh(x) = \delta(a) T_m^{(r)}(a) \\ &+ \delta(-a) T_m^{(r)}(-a) + O(m^r g^m(a-\varepsilon)) \\ &= m^r g^m(a)(a^2-1)^{-\frac{r}{2}}\left\{\delta(a) + (-1)^{m+r}\delta(-a) + O\left(\frac{1}{m}\right)\right\},\end{aligned}$$

where $\delta(\pm a)$ are the steps of the function $h(x)$ at the points $x = \pm a$. Selecting m equal to n or $n-1$ such that the numbers $\delta(a)$ and $(-1)^{m+r}\delta(-a)$ have one sign, we obtain

$$\begin{aligned}&\|\Psi_a^{(r)}\|_{n,p} \geqslant \|\Psi_a^{(r)}\|_n \geqslant |\Psi_a^{(r)}(T_m)| \\ &= (a^2-1)^{-\frac{r}{2}} m^r g^m(a)\left\{|\delta(a)| + |\delta(-a)| + O\left(\frac{1}{n}\right)\right\} > C n^r g^n(a) \quad (C > 0),\end{aligned}$$

which coincides with the first inequality (3.6).

Let us estimate the norm of the functional $\Psi_{\alpha,a}^{(r)}$.

LEMMA 3.2. *Let $r \geqq 0$ be an integer, and take $a > 1$, $\alpha > -1$. For any functional $\Psi_{\alpha,a}^{(r)}$ there are numbers $0 < \nu_1 \leqq \nu_2 < \infty$ such that for all $n = 1, 2, \cdots$ and all p $(1 \leqq p \leqq \infty)$*

$$\nu_1 n^{r-1-\alpha} g^n(a) \leqslant \|\Psi_{\alpha,a}^{(r)}\|_{n,p} \leqslant \nu_2 n^{r-1-\alpha} g^n(a)$$

PROOF. By analogy with the proof of Lemma 3 of [8] we can show that when $n \to \infty$

$$\int_1^a g^n(x)(a^2-x^2)^\alpha \varphi(x)\, dx = C_1(\alpha, a)\frac{g^n(a)}{n^{\alpha+1}}\{\varphi(a) + o(1)\} \tag{3.7}$$

and

$$\int_{-a}^{-1} g^n(x)(a^2-x^2)^\alpha \varphi(x)\, dx = C_1(\alpha, a,)\frac{g^n(a)}{n^{\alpha+1}}\{(-1)^n \varphi(-a) + o(1)\}. \tag{3.8}$$

Therefore, integrating (3.4) with the weight $(a^2 - x^2)^\alpha |\varphi(x)|$, just as when proving Lemma 3.1 we obtain

$$\|\Psi'^{(r)}_{\alpha, a}\|_{n, p} \leqslant \nu_2 n^{r-1-\alpha} g^n(a) \quad (1 \leqslant p \leqslant \infty).$$

We estimate the lower bound of $\| \Psi^{(r)}_{\alpha,a} \|_{n,p}$ as follows. We define

$$p^*_n(x) = \begin{cases} T_n(x). \text{ if } & \varphi(a)\varphi(-a) \geqslant 0, \quad r+n = 2m \\ & \text{or } \varphi(a)\varphi(-a) \leqslant 0, \quad r+n = 2m+1; \\ T_{n-1}(x), \text{ if } & \varphi(a)\varphi(-a) > 0, \quad r+n = 2m+1 \\ & \text{or } \varphi(a)\varphi(-a) < 0, \quad r+n = 2m. \end{cases}$$

Applying (3.7) and (3.8), we obtain

$$\begin{aligned} \|\Psi'^{(r)}_{\alpha, a}\|_{n, p} &\geqslant \|\Psi'^{(r)}_{\alpha, a}\|_{n} \geqslant |\Psi'^{(r)}_{\alpha, a}(p^*_n)| \\ &= C_1(\alpha, a) \frac{g^n(a)}{n^{\alpha+1-r}} \{|\varphi(a)| + |\varphi(-a)| + o(1)\} > \nu_1 g^n(a) n^{r-1-\alpha}. \end{aligned}$$

The lemma is proved.

Let $f(x) \in L_p[-1,1]$ $(1 \leqq p \leqq \infty)$ and let the sequence of polynomials $\{\hat{p}_n(x;f)\}$ be such that

$$\|f - \hat{p}_n(f)\|_{L_p[-1,1]} \leqslant F_n,$$

where $F_n \downarrow 0$. The question is how the sequences $\{\Psi_{a,r}(\hat{p}_n(f))\}$ behave when $n \to \infty$. We shall answer this question below in the case where nothing more is known about the function $f(x)$.

Let us remember that we denoted by $p_n(f,p)$ the polynomials satisfying the equality

$$E_n(f)_p = \|f - p_n(f, p)\|_{L_p[-1,1]},$$

and that

$$p_n(x; f) = p_n(x; f, \infty).$$

In Lemma 3.3 and its Corollary 3.2 below we impose certain restrictions on the form of the homogeneous additive functional Ψ, and we restrict only the order of the norms $\| \Psi \|_{n,p}$. We can take as Ψ, in particular, the functional $\Psi_{a,r}$ or $\Psi^{(r)}_{\alpha,a}$.

Lemma 3.3. *Let ν, $a > 1$, $1 \leqq p \leqq \infty$, $0 < \epsilon < 1$ be given real numbers, let $k \geqq 0$ be a fixed integer and let $\Psi(p_n)$ $(n = 0, 1, \cdots)$ be any additive homogeneous functional satisfying the conditions*

$$\|\Psi\|_{n, p} = O(n^\nu g^n(a)) \quad (n \to \infty).$$

Then for any function $f(x) \in L_p[-1,1]$ one has

$$\begin{aligned} \Psi(p_n(f, p)) &= \Psi(p_k(f, p)) \\ &+ O\{n^\nu [g^{\epsilon n}(a) E_k(f)_p + g^n(a) E_{[\epsilon n]+1}(f)_p]\}, \end{aligned} \tag{3.9}$$

where the constant entering into O does not depend on $f(x)$.

If $f(x) \equiv \dot{p}_k(x)$ formula (3.9) becomes trivial, and in the remaining cases it is the upper bound of $\Psi(p_n(f,p))$.

PROOF. Since

$$p_n(x;f,p) \equiv p_k(x;f,p) + \sum_{l=k+1}^{n} [p_l(x;f,p) - p_{l-1}(x;f,p)],$$

we have

$$|\Psi(p_n(f,p)) - \Psi(p_k(f,p))| \leqslant 2 \sum_{l=k+1}^{n} \|\Psi\|_{l,p} E_{l-1}(f)_p \leqslant$$
$$C(\Psi)\Big\{E_k(f)_p \sum_{l=k+1}^{[\varepsilon n]+1} l^\nu g^l(a) + E_{[\varepsilon n]+1}(f)_p \sum_{l=[\varepsilon n]+2}^{n} l^\nu g^l(a)\Big\}.$$

Taking into account the fact that the number ν may be negative, let us continue the argument. The sequence $\{l^\nu g^{l/2}(a)\}$ for $l > l_0 \geqq -(\nu/2)\ln g(a)$ is increasing. Therefore

$$\sum_{l=k+1}^{[\varepsilon n]+1} l^\nu g^l(a) \leqslant \sum_{l=1}^{l_0} l^\nu g^l(a) + ([\varepsilon n]+1)^\nu g^{\frac{[\varepsilon n]+1}{2}}(a) \sum_{l=l_0}^{[\varepsilon n]+1} g^{\frac{l}{2}}(a)$$
$$\leqslant C + C([\varepsilon n]+1)^\nu g^{[\varepsilon n]+1}(a) = O(n^\nu g^{\varepsilon n}(a)).$$

Analogously

$$\sum_{l=[\varepsilon n]+2}^{n} l^\nu g^l(a) = O(n^\nu g^n(a)),$$

and the lemma is proved.

COROLLARY 3.2. *Let the functional Ψ and the parameters ν, a, p, ϵ be the same as in Lemma 3.3, and let $\{F_n\}$ be the given sequence of real numbers. If $f(x) \in L_p[-1,1]$ and the polynomials $\hat{p}_n(f)$ are such that*

$$\|f(x) - \hat{p}_n(x;f)\|_{L_p[-1,1]} \leqslant F_n \qquad (n = 1, 2, \ldots),$$

then

$$\Psi(\hat{p}_n(f)) = \Psi(p_k(f,p)) + O\{n^\nu [F_k g^{\varepsilon n}(a) + g^n(a)(F_n + F_{[\varepsilon n]+1})]\},$$

where the constant entering into O does not depend on the function $f(x)$ and the sequence $\{F_n\}$.

This corollary follows directly from Lemma 3.3, since

$$\Psi(\hat{p}_n(f)) = \Psi(p_n(f,p)) + O\{\|\Psi\|_{n,p}(F_n + E_n(f)_p)\}.$$

Let us now consider how the sequence $\{\Psi(\hat{p}_n(f))\}$ behaves in the case when $f(z) \in MH(k,\beta,p,G(b))$. First, let us again consider the case where $\hat{p}_n(f) = p_n(f,p')$. Let $\delta_{a,b}$ be the Kronecker symbol.

THEOREM 3.1. *Let $k \geqq 0$ be an integer, $0 < \beta \leqq 1$ when $k > 0$, $0 \leqq \beta \leqq 1$ when $k = 0$, let $a > 1$, $b > 1$, $1 \leqq p' \leqq \infty$, $1 \leqq p \leqq \infty$, $m > 0$, and let ν be a real number. Let $\Psi(p_n)$ $(n = 0, 1, \cdots)$ be any additive homogeneous functional satisfying the condition*

$$\|\Psi\|_{n,p'} = O(n^{\nu} g^n(a)) \quad (n \to \infty).$$

For any function $f(z) \in MH(k, \beta, p, G(b))$ *set*

$$\Phi(f) = \begin{cases} \lim\limits_{n\to\infty} \Psi(p_n(f, p')), & \text{if } a = b, \quad k - \nu + \beta > 1 \\ & \text{or } a < b; \\ \Psi(p_k(f, p')), & \text{if } a = b, \quad k - \nu + \beta \leqslant 1 \\ & \text{or } a > b. \end{cases}$$

Then

$$\Psi(p_n(f, p')) = \Phi(f) + O\left\{\frac{g^n(a)}{g^n(b)}[n^{\nu-k-\beta} + \delta_{a,b}(n^{\nu+1-k-\beta} + \delta_{k+\beta,\nu+1}\ln n)]\right\}$$
$$(n \to \infty),$$

where the constant entering into O *depends only on the parameters* M, K, β, p, b *and on the functional* Ψ.

Depending on the ratio of the parameters a, b, ν, k, β this formula is asymptotic for the individual function $f(x)$ or gives the upper bound of the growth of the sequence $\{\Psi(p_n(f,p'))\}_n$. The formulas of all the remaining theorems in this section have the same structure.

PROOF. Let us consider the series

$$\Psi(p_n(f, p')) + \sum_{l=n+1}^{\infty} \Psi(p_l(f, p') - p_{l-1}(f, p')). \tag{3.10}$$

By Theorem 2.1 it is majorized by the series

$$\sum_{l=n+1}^{\infty} l^{\nu-k-\beta} g^l(a) g^{-l}(b).$$

If $a < b$ or if $a = b$ and $k - \nu + \beta > 1$ this series converges; therefore the initial series also converges. Since the partial sums of the series (3.10) have the form $\Psi(p_N(f,p'))$, its sum equals

$$\Phi(f) = \lim_{N\to\infty} \Psi(p_N(f, p')),$$

and this limit exists. Thus for the indicated values of a, b, k, ν, p and β we have

$$|\Psi(p_n(f, p') - \Phi(f))| \leqslant C(M, \Psi) \sum_{l=n+1}^{\infty} l^{\nu-k-\beta} \frac{g^l(a)}{g^l(b)}$$
$$\leqslant \begin{cases} C(M, \Psi)\, n^{\nu-k-\beta} g^n(a) g^{-n}(b), & \text{if } a < b, \\ C(M, \Psi)\, n^{\nu+1-k-\beta}, & \text{if } a = b, \quad k - \nu + \beta > 1. \end{cases}$$

For the remaining values of the parameters we have

$$|\Psi(p_n(f, p')) - \Psi(p_k(f, p'))| = |\sum_{l=k+1}^{n} [\Psi(p_l(f, p'))$$
$$- \Psi(p_{l-1}(f, p'))]| \leqslant C(\Psi, M) \sum_{l=k+1}^{n} l^{\nu-k-\beta} \frac{g^l(a)}{g^l(b)}$$
$$\leqslant \begin{cases} C(\Psi, M)\, n^{\nu-k-\beta} g^n(a)\, g^{-n}(b), & \text{if} \quad a > b, \\ C(\Psi, M)\, n^{\nu+1-k-\beta}, & \text{if} \quad a = b, \quad k+\beta-\nu < 1, \\ C(\Psi, M) \ln n, & \text{if} \quad a = b, \quad k+\beta-\nu = 1. \end{cases}$$

The theorem is proved.

COROLLARY 3.3. *Let the parameters k, β, p, p', a, b, ν, M and the functionals $\Psi(p_n)$ and $\Phi(f)$ be the same as in Theorem 3.1, and let $\{F_n\}$ be the given sequence of positive numbers. If $f(z) \in MH(k, \beta, p, G(b))$, and the polynomials $\hat{p}_n(f)$ are such that*

$$\|f(x) - \hat{p}_n(x; f)\|_{L_{p'}[-1,1]} \leqslant F_n \; (n = k, k+1, \ldots),$$

then when $n \to \infty$ we get

$$\Psi(\hat{p}_n(f)) = \Phi(f) + O\Big\{F_n n^\nu g^n(a) + \frac{g^n(a)}{g^n(b)}[n^{\nu-k-\beta} + \delta_{a,b}(n^{\nu+1-k-\beta} + \delta_{k+1,\nu+1} \ln n)]\Big\}, \tag{3.11}$$

where the constant entering into O depends only on the parameters M, k, β, p, b and on the functional Ψ.

In fact, we have

$$|\Psi(\hat{p}_n(f)) - \Psi(p_n(f, p'))| \leqslant \|\Psi\|_{n,p'} \|\hat{p}_n(f) - p_n(f, p')\|_{L_{p'}[-1,1]}$$
$$\leqslant C(\Psi, M)\, n^\nu g^n(a) \{F_n + n^{-k-\beta} g^{-n}(b)\},$$

and the statements of the corollary follow from Theorem 3.1.

If formula (3.11) is not asymptotic we can say nothing of its accuracy in the sense of order when $n \to \infty$. For the functionals $\Psi_{a,r}$ it obviously is rough. In any case we shall present more exact formulas below for the functionals $\Psi_a^{(r)}$ and $\Psi_{\alpha,a}^{(r)}$.

THEOREM 3.2. *Let $r \geqq 0$, $k \geqq 0$ be integers, $a > 1$, $b > 1$, $0 < \beta \leqq 1$, $M > 0$, and let $f \in MH(k, \beta, \infty, G(b))$. Set*

$$\Phi_a^{(r)}(f) = \begin{cases} \displaystyle\int_{-a}^{a} f^{(r)}(x)\, dh(x), & \text{if} \quad a = b \quad r \leqslant k \\ & \text{or} \quad a < b; \\ \Psi_a^{(r)}(p_k(f)), & \text{if} \quad a = b \quad r > k \\ & \text{or} \quad a > b, \end{cases}$$

where $h(x)$ is a function defining the functional $\Psi_a^{(r)}$. Then when $n\to\infty$ one has

$$\Psi_a^{(r)}(p_n(f))=\Phi_a^{(r)}(f)+O\left\{n^{r-k-\beta}(1+\delta_{a,b}\ln n)\frac{g^n(a)}{g^n(b)}\right\},\tag{3.12}$$

where the constant entering into O depends only on the numbers M, k, β, p and on the functional $\Psi_a^{(r)}$.

The proof will be divided into several parts.

1) Let $a<b$. Then

$$\Psi_a^{(r)}(p_n(f))=\Phi_a^{(r)}(f)+\int_{-a}^{a}[p_n^{(r)}(x;f)-f^{(r)}(x)]\,dh(x).$$

Since for $x\in\overline{G(a)}$

$$f^{(r)}(x)-p_n^{(r)}(x;f)=\sum_{l=n}^{\infty}[p_{l+1}^{(r)}(x;f)-p_l^{(r)}(x;f)],$$

then from inequality (3.5), the inequality

$$|p_n^{(r)}(\xi)|\leqslant n^{2r}\|p_n(x)\|_{C[-1,1]}\qquad(\xi\in[-1,1])$$

of A. A. Markov [**19**], and inequality (2.5) we obtain

$$|f^{(r)}(x)-p_n^{(r)}(x;f)|\leqslant C\sum_{l=n}^{\infty}l^r g^l(a)\|p_{l+1}(\xi,f)-p_l(\xi,f)\|_{C[-1,1]}$$

$$\leqslant C(M)\sum_{l=n}^{\infty}l^r g^l(a)E_l(f)\leqslant C(M)\sum_{l=n}^{\infty}l^{r-k-\beta}\frac{g^l(a)}{g^l(b)}\leqslant C(M)n^{r-k-\beta}g^n(a)g^{-n}(b).$$

Consequently

$$\left|\int_{-a}^{a}\{p_n^{(r)}(x;f)-f^{(r)}(x)\}\,dh(x)\right|\leqslant C(M,\Psi_a^{(r)})\,n^{r-k-\beta}\frac{g^n(a)}{g^n(b)},$$

and equality (3.12) is proved in case $a<b$.

2) Now let $a=b$ and $r\leqq k$. Let us select $a_0<a$ such that the function $h(x)$ is constant in the intervals $(-a,-a_0)$, (a_0,a), and let us represent the functional $\Psi_a^{(r)}(p_n)$ in the form

$$\begin{aligned}\Psi_a^{(r)}(p_n)&=\int_{-a_0}^{a_0}p_n^{(r)}(x)\,dh(x)+\delta(a)\,p_n^{(r)}(a)+\delta(-a)\,p_n^{(r)}(-a)\\&=\Psi_{a_0}^{(r)}(p_n)+\delta(a)\,F_r(p_n)+\delta(-a)\,G_r(p_n),\end{aligned}\tag{3.13}$$

where $\delta(\pm a)$ are the steps of $h(x)$ at the point $x=\pm a$. In accord with the preceding, we get

$$\Psi_{a_0}^{(r)}(p_n)=\int_{-a_0}^{a_0}f^{(r)}(x)\,dh(x)+O\left(n^{r-k-\beta}\frac{g^n(a_0)}{g^n(a)}\right).\tag{3.14}$$

The values of the functionals $F_r(p_n(f))$ and $G_r(p_n(f))$ are estimated identically, and we shall evaluate only $F_r(p_n(f))$. For this purpose let us extend the functional $F_r(p_n)$ from the subspace of the polynomials of degree

n to the subspace $H(k, \beta, \infty, G(a))$ according to the formula

$$F_{r,n}(f) = \frac{1}{\pi}\int_{-1}^{1} f^{(r)}(\xi)\sin(n+1)\arccos\xi\, \frac{g^{n+1}(a)}{a-\xi}\, d\xi \tag{3.15}$$

Since (see Lemma 1 in [8])

$$\frac{1}{\pi}\int_{-1}^{1} p_n^{(r)}(\xi)\sin(n+1)\arccos\xi\,\frac{g^{n+1}(a)}{a-\xi}\,d\xi = p_n^{(r)}(a),$$

we get $F_{r,n}(p_n) = F_r(p_n)$, and consequently

$$F_r(p_n(f)) = F_{r,n}(p_n(f) - f) + F_{r,n}(f). \tag{3.16}$$

Therefore, in order to estimate $F_r(p_n(f))$ it is sufficient to evaluate the components in the right-hand side of equation (3.16).

By analogy with Lemma 1 of [8], applying the Cauchy theorem, we have

$$\begin{aligned} F_{r,n}(p_n(f) - f) &= \frac{1}{\pi}\int_{-1}^{1}[p_n^{(r)}(\xi; f) - f^{(r)}(\xi)]\sin(n+1)\arccos\xi\,\frac{g^n(a)}{a-\xi}\,d\xi \\ &= \frac{1}{2\pi i}\oint_{\Gamma_{a_0}}[p_n^{(r)}(\lambda; f) - f^{(r)}(\lambda)]\frac{g^n(a)}{\lambda - a}\frac{d\lambda}{g^n(\lambda)} \\ &= \frac{1}{2\pi i}\sum_{l=n}^{\infty}\int_{\Gamma_{a_0}}[p_{l+1}^{(r)}(\lambda; f) - p_l^{(r)}(\lambda; f)]\frac{g^n(a)}{g^n(\lambda)}\frac{d\lambda}{\lambda - a} \\ &= \frac{1}{2\pi i}\sum_{l=n}^{\infty}\int_{\Gamma_{a_0}}\frac{1}{\pi}\int_{-1}^{1}[p_{l+1}(\xi; f) - p_l(\xi; f)]\sin(l+2)\arccos\xi \\ &\times \frac{d^r}{d\lambda^r}\left(\frac{g^{l+2}(\lambda)}{\lambda - \xi}\right)d\xi\,\frac{g^n(a)}{g^n(\lambda)}\frac{d\lambda}{\lambda - a}. \end{aligned}$$

From this we obtain

$$\begin{aligned} |F_{r,n}(p_n(f) - f)| &\leqslant C\sum_{l=n}^{\infty}\frac{g^n(a)}{g^n(a_0)}l^r g^l(a_0)\|p_{l+1}(\xi; f) - p_l(\xi; f)\|_{C[-1,1]} \\ &\leqslant C(M)\frac{g^n(a)}{g^n(a_0)}\sum_{l=n}^{\infty}l^{r-k-\beta}\frac{g^l(a_0)}{g^l(a)} \leqslant C(M)\,n^{r-k-\beta}. \end{aligned} \tag{3.17}$$

To evaluate $F_{r,n}(f)$ let us assume $a_n = a - (1/n)$. As in the preceding case we have

$$\begin{aligned} F_{r,n}(f) &= \frac{1}{\pi}\int_{-1}^{1} f^{(r)}(\xi)\sin(n+1)\arccos\xi\,\frac{g^{n+1}(a)}{a-\xi}\,d\xi \\ &= \frac{1}{2\pi i}\int_{\Gamma_{a_n}}\frac{f^{(r)}(\lambda)}{g^{n+1}(\lambda)}\frac{g^{n+1}(a)}{\lambda - a}\,d\lambda \\ &= \frac{1}{2\pi i}\int_{\Gamma_{a_n}}\frac{f^{(r)}(\lambda) - p_n(\lambda) - f^{(r)}(a) + p_n(a)}{g^{n+1}(\lambda)}\frac{g^{n+1}(a)}{\lambda - a}\,d\lambda \\ &+ f^{(r)}(a) = f^{(r)}(a) + J_1. \end{aligned} \tag{3.18}$$

where $p_n(\lambda)$ is any polynomial of degree n. Since $f(\lambda) \in MH(k, \beta, \infty, G(a))$, $r \leqq k$, we have $f^{(r)}(\lambda) \in MH(k-r, \beta, \infty, G(a))$. Applying inequality (2.9) for this case, we obtain

$$|J_1| \leqslant \frac{1}{\pi} \inf_{p_n} \max_{\lambda \in \Gamma_{a_n}} |f^{(r)}(\lambda) - p_n(\lambda) - f^{(r)}(a) - p_n(a)| \frac{g^{n+1}(a)}{g^{n+1}(a_n)} \int_{\Gamma_{a_n}} \frac{|d\lambda|}{|\lambda - a|}$$
$$\leqslant C \ln n \frac{g^{n+1}(a)}{g^{n+1}(a_n)} \inf_{p_n} \max_{\lambda \in \Gamma_a} |f^{(r)}(\lambda) - p_n(\lambda)| \leqslant C(M) n^{r-k-\beta} \ln n,$$

that is, $F_{r,n}(f) = f^{(r)}(a) + O(n^{r-k-\beta} \ln n)$.

Analogously, $G_r(p_n(f)) = f^{(r)}(-a) + O(n^{r-k-\beta} \ln n)$.

From all these evaluations it follows that

$$\begin{aligned} \Psi_a^{(r)}(p_n(f)) &= \Phi_{a_0}^{(r)}(f) + \delta(a) f^{(r)}(a) + \delta(-a) f^{(r)}(-a) \\ &+ O(n^{r-k-\beta} \ln n) = \Phi_a^{(r)}(f) + O(n^{r-k-\beta} \ln n), \end{aligned} \tag{3.19}$$

which proves formula (3.12) in the case where $a = b$ and $r \leqq k$.

3) Let us consider the case $a = b$ and $r > k$. Retaining the representation (3.13) of the functional $\Psi_a^{(r)}(p_n)$, we now continue the functional $F_r(p_n)$ in a different way to $H(k, \beta, \infty, G(a))$, namely, we assume that

$$F_{r,n}(f) = \frac{1}{\pi} \int_{-1}^{1} f^{(k)}(\xi) \sin(n+1) \arccos \xi \frac{d^{r-k}}{da^{r-k}} \left(\frac{g^{n+1}(a)}{a - \xi} \right) d\xi. \tag{3.20}$$

We again have

$$F_r(p_n(f)) = F_{r,n}(p_n(f) - f) + F_{r,n}(f), \tag{3.16'}$$

and, estimating as in the proof of inequality (3.17), we obtain

$$|F_{r,n}(p_n(f) - f)| \leqslant C(M) n^{r-k-\beta} \tag{3.21}$$

Let us present the estimate of $F_{r,n}(p_n(f))$ considering two separate cases.

a) Let $0 < \beta < 1$. Since (see Lemma 1 in [8])

$$\int_{-1}^{1} \sin(n+1) \arccos \xi \frac{d^{r-k}}{da^{r-k}} \left(\frac{g^{n+1}(a)}{a - \xi} \right) d\xi = 0,$$

we have

$$\begin{aligned} F_{r,n}(f) &= \frac{1}{\pi} \int_{-1}^{1} [f^{(k)}(\xi) - f^{(k)}(a)] \sin(n+1) \arccos \xi \frac{d^{r-k}}{da^{r-k}} \left(\frac{g^{n+1}(a)}{a - \xi} \right) d\xi \\ &= \frac{1}{2\pi i} \int_{\Gamma_{a_n}} \frac{f^{(k)}(\lambda) - f^{(k)}(a)}{g^{n+1}(\lambda)} \frac{d^{r-k}}{da^{r-k}} \left(\frac{g^{n+1}(a)}{\lambda - a} \right) d\lambda \\ &= \sum_{l=1}^{r-k} (n+1)^{r-k-l} g^{n+1}(a) C_l(a) \int_{\Gamma_{a_n}} \frac{f^{(k)}(\lambda) - f^{(k)}(a)}{g^{n+1}(\lambda)} \frac{d\lambda}{(\lambda - a)^{l+1}} \\ &+ C(a)(n+1)^{r-k} g^{n+1}(a) \int_{\Gamma_{a_n}} \frac{f^{(k)}(\lambda) - f^{(k)}(a)}{g^{n+1}(\lambda)} \frac{d\lambda}{\lambda - a} = J_2 + J_3. \end{aligned}$$

From the conditions $f \in MH(k, \beta, \infty, G(a))$ it follows that when $0 < \beta < 1$

$$| f^{(k)}(\lambda) - f^{(k)}(a) | \leqslant M | \lambda - a |^{\beta}.$$

Therefore

$$| J_2 | \leqslant \sum_{l=1}^{r-k} C(M)(n+1)^{r-k-l} \frac{g^{n+1}(a)}{g^{n+1}(a_n)} \int_{\Gamma_{a_n}} \frac{|d\lambda|}{|\lambda - a|^{l+1-\beta}}$$

$$\leqslant \sum_{l=1}^{r-k} C(M) n^{r-k-l} | a_n - a |^{\beta - l} \leqslant C(M) n^{r-k-\beta}.$$

For the integral J_3 we have

$$| J_3 | = \left| C(a)(n+1)^{r-k} g^n(a) \int_{\Gamma_{a_n}} \frac{f^{(k)}(\lambda) - p_n(\lambda) - f^{(k)}(a) + p_n(a)}{g^{n+1}(\lambda)} \frac{d\lambda}{\lambda - a} \right|$$

$$\leqslant C(M)(n+1)^{r-k-\beta} \frac{g^{n+1}(a)}{g^{n+1}(a_n)} \int_{\Gamma_{a_n}} \frac{|d\lambda|}{|\lambda - a|} \leqslant$$

$$\leqslant C(M) n^{r-k-\beta} | \ln(a - a_n) | \leqslant C(M) n^{r-k-\beta} \ln n,$$

since $p_n(\lambda)$ may be selected in such a way that $\max_{\lambda \in \Gamma_{a_n}} | f^{(k)}(\lambda) - p_n(\lambda) |$ is minimized.

From the estimate of J_2 and J_3 we find that when $0 < \beta < 1$

$$F_{r,n}(f) = O(n^{r-k-\beta} \ln n).$$

b) $F_{r,n}(f)$ is evaluated in a somewhat more complex manner in the case where $\beta = 1$. It is easy to prove that

$$F_{r,n}(f) = \frac{1}{2\pi i} \int_{\Gamma_{a_n}} \frac{f^{(k)}(\lambda) - \pi_n(\lambda; f^{(k)})}{g^{n+1}(\lambda)} \frac{d^{r-k}}{da^{r-k}} \left(\frac{g^{n+1}(a)}{\lambda - a} \right) d\lambda + \pi_n^{(r-k)}(a; f^{(k)}),$$

where $\pi_n(\lambda; f^{(k)})$ is defined by the equality

$$\inf_{p_n} \max_{\lambda \in \Gamma_a} | f^{(k)}(\lambda) - p_n(\lambda) | + \max_{\lambda \in \Gamma_a} | f^{(k)}(\lambda) - \pi_n(\lambda; f^{(k)}) |.$$

On the basis of the inequality (2.9) we get

$$| f^{(k)}(\lambda) - \pi_n(\lambda; f^{(k)}) \leqslant C(M)(n+1)^{-1} \quad (n > 0, \ \lambda \in \Gamma_a).$$

Then, assuming that $n_{-1} = 0$, $n_0 = 1$, $n_1 = 2, \cdots, n_l = 2^l, \cdots, n_{m-1} = 2^{m-1} < n_m = n \leqq 2^m$, we obtain

$$\pi_n^{(r-k)}(a; f^{(k)}) = \pi_0^{(r-k)}(a; f^{(k)}) + \sum_{l=0}^{m} [\pi_{n_l}^{(r-k)}(a; f^{(k)}) - \pi_{n_{l-1}}(a; f^{(k)})].$$

Since (see [**13**], p. 357)

$$| p_n'(\lambda) | \leqslant Cn \max_{\xi \in \Gamma_a} | p_n(\xi) | \quad (\lambda \in \Gamma_a), \tag{3.22}$$

we get

$$|\pi_n^{(r-k)}(a; f^{(k)})| \leqslant 2C \sum_{l=0}^{m} n_l^{r-k} \max_{\lambda\in\Gamma_a} |f^{(k)}(\lambda) - \pi_{n_{l-1}}(\lambda; f^{(k)})|$$

$$\leqslant C(M) \sum_{l=0}^{m} 2^{l(r-k-1)} \leqslant C(M)\, m 2^{m(r-k-1)} \leqslant C(M)\, n^{r-k-1} \ln n.$$

Consequently

$$F_{r,n}(f) \leqslant \frac{C(M)}{n} \frac{g^{n+1}(a)}{g^{n+1}(a_n)} \sum_{l=0}^{r-k} n^{r-k-l} \int_{\Gamma_{a_n}} \frac{|d\lambda|}{|\lambda - a|^{l+1}}$$

$$+ |\pi_n^{(r-k)}(a; f^{(k)})| \leqslant C(M)\, n^{r-k-1} \ln n \ (r > k, \quad \beta = 1),$$

i.e. when $\beta = 1$ we also have

$$F_{r,n}(f) = O(n^{r-k-\beta} \ln n).$$

Combining this estimate $(0 < \beta \leqq 1)$ with formulas (3.13), (3.16′) and (3.21), we obtain

$$|\Psi_a^{(r)}(p_n(f))| \leqslant \left| \int_{-a_0} f^{(r)}(x)\, dh(x) \right| + O(n^{r-k-\beta} \ln n).$$

Since $a_0 < a$ and $r > k$,

$$\left| \int_{-a_0}^{a_0} f^{(r)}(x)\, dh(x) \right| \leqslant C \sum_{l=k}^{\infty} \| p_{l+1}^{(r)}(x; f) - p_l^{(r)}(x; f) \|_{C_{[-a_0, a_0]}}$$

$$\leqslant C \sum_{l=k}^{\infty} l^r g^l(a_0) \| p_{l+1}(x; f) - p_l(x; f) \|_{C_{[-1, 1]}} \tag{3.23}$$

$$\leqslant C(M) \sum_{l=k}^{\infty} l^{r-k-\beta} g^l(a) g^{-l}(a) \leqslant C(M)$$

and consequently

$$|\Psi_a^{(r)}(p_n(f))| \leqslant C(M, \Psi_a^{(r)})\, n^{r-k-\beta} \ln n \quad (r > k, 0 < \beta \leqslant 1),$$

which coincides in case 3) with the formula (3.12).

4) Finally, let $f(z) \in MH(k, \beta, \infty, G(b))$ and $a > b$. It would be possible in this case to present arguments using the Cauchy formula, but there is a simpler proof. We have

$$\Psi_a^{(r)}(p_n(f)) = \Psi_a^{(r)}(p_k(f)) + \sum_{l=k+1}^{n} \Psi_a^{(r)}(p_l(f) - p_{l-1}(f))$$

and

$$\left| \sum_{l=k+1}^{n} \Psi_a^{(r)}(p_l(f) - p_{l-1}(f)) \right| \leqslant 2 \sum_{l=k+1}^{n} \| \Psi_a^{(r)} \|_l E_{l-1}(f)$$

$$\leqslant C(M, \Psi_a^{(r)}) \sum_{l=k+1}^{n} l^{r-k-\beta} \frac{g^l(a)}{g^l(b)} = O\left(n^{r-k-\beta} \frac{g^n(a)}{g^n(b)} \right).$$

Theorem 3.2 is completely proved. The following corollary comes from it.

COROLLARY 3.4. *Let* $\{F_n\}$ *be the given sequence of positive numbers, let* r *and* k *be nonnegative integers, and let* $a > 1$, $b > 1$, $0 < \beta \leqq 1$, $M > 0$, $1 \leqq p' \leqq \infty$. *If for the function* $f(z) \in MH(k, \beta, \infty, G(b))$ *there is a sequence of polynomials* $\{\hat{p}_n(f)\}$ *satisfying the inequality*

$$\|f(x) - \hat{p}_n(x; f)\|_{L_{p'}[-1, 1]} \leqslant F_n \qquad (n = 1, 2, \ldots),$$

then for the sequence $\{\Psi_a^{(r)}(\hat{p}_n(f))\}$ *when* $n \to \infty$ *the following formula holds:*

$$\Psi_a^{(r)}(\hat{p}_n(f)) = \Phi_a^{(r)}(f) + O\left\{n^r g^n(a)\left(F_n + \frac{1 + \delta_{a, b} \ln n}{n^{k+\beta} g^n(b)}\right)\right\}, \tag{3.24}$$

where $\Phi_a^{(r)}(f)$ *has the same value as in Theorem* 3.2, *and the constant entering into* O *depends only on the numbers* M, k, β, b *and on the functional* $\Psi_a^{(r)}$.

In fact, on the basis of Corollary 3.1,

$$\begin{aligned}\Psi_a^{(r)}(\hat{p}_n(f)) &= \Psi_a^{(r)}(p_n(f)) + \Psi_a^{(r)}(\hat{p}_n(f) - p_n(f)) \\ &= \Psi_a^{(r)}(p_n(f)) + O\{\|\Psi_a^{(r)}\|_{n,p}(\|\hat{p}_n(f) - \|_{L_{p'}} + \|f - p_n(f)\|_C)\} \\ &= \Psi_a^{(r)}(p_n(f)) + O\{n^r g^n(a)(F_n + n^{-k-\beta} g^{-n}(b))\}.\end{aligned}$$

From this and from Theorem 3.2 we have the statement of the corollary.

REMARKS. 1. If under the conditions of Corollary 3.4

$$F_n = O(n^{-k-1-\beta_1} g^{-n}(b_1)) \quad (\beta < \beta_1, b \leqslant b_1),$$

the estimate of the "remainder term" in formula (3.24) is quite rough. However, in this case, on the basis of the inverse theorem ([**15**], 1951 ed., p. 367) there exists a number $M_1 > 0$ such that $f(z) \in M_1 H(k + [\beta_1], \beta_1 - [\beta_1], \infty, G(b_1))$. Applying formula (3.24) to this class, we can obtain a more exact upper bound for $|\Psi_a^{(r)}(\hat{p}_n(f)) - \Phi_a^{(r)}(f)|$.

2. If under the conditions of Corollary 3.4 we have $a < b$ and

$$F_n = \varphi_n n^{-k-\beta} g^{-n}(b) \quad (\varphi_n \downarrow 0, n \to \infty),$$

then, repeating the corresponding part of the proof of Theorem 3.2, for the sequence $\Psi_a^{(r)}(\hat{p}_n(f))$ it is possible to obtain a more exact asymptotic formula than (3.24), namely

$$\Psi_a^{(r)}(\hat{p}_n(f)) = \Phi_a^{(r)}(f) + O(n^r g^n(a) F_n). \tag{3.25}$$

When $a \geqq b$ our method of proof does not permit us to obtain analogous refinement of formula (3.24).

3. As will be obvious from the results of §4 (see Theorem 4.3), in the general case in the class $MH(k, \beta, \infty, G(b))$ the order of the "remainder term" in formula (3.24) cannot be reduced.

These remarks remain valid with obvious changes for all of the results presented below with respect to the sequences of polynomials converging to analytic functions, and we shall not repeat them.

Let us now consider how the sequence $\{\Psi_a^{(r)}(p_n(f))\}$ behaves when $f(z) \in MH(k, \beta, p, G(b))$ $(1 \leq p < \infty)$.

THEOREM 3.3. *Let r and k be nonnegative integers, let* $a > 1$, $b > 1$, $0 < \beta \leq 1$, $1 \leq p < \infty$, $M > 0$, *and let* $f(z) \in MH(k, \beta, p, G(b))$. *Set*[4]

$$\Phi_a^{(r)}(f) = \begin{cases} \int_{-a}^{a} f^{(r)}(x)\, dh(x), & \text{if } a = b, \quad r + \frac{1}{p} < k + \beta \\ & \text{or } a < b; \\ \Psi_a^{(r)}(p_k(f)), & \text{if } a = b, \quad r + \frac{1}{p} \geqslant k + \beta \\ & \text{or } a > b, \end{cases}$$

where $h(x)$ is the function defining the functional $\Psi_a^{(r)}$. Then, for the sequence $\{\Psi_a^{(r)}(p_n(f))\}$ when $n \to \infty$,

$$\Psi_a^{(r)}(p_n(f)) = \Phi_a^{(r)}(f) + O\left\{n^{r-k-\beta}\left[1 + \delta_{a,b}\, n^{1/p}(1 + \delta_{r+\frac{1}{p},\, k+\beta} \ln n)\right]\frac{g^n(a)}{g^n(b)}\right\}. \tag{3.26}$$

The constant entering into O depends only on the numbers M, k, β, p, b and the functional $\Psi_a^{(r)}$.

PROOF. 1) If $a \neq b$, the proof differs in no way from the proof of Theorem 3.2. Therefore let $a = b$ and at first let $k + \beta > r + 1/p$. Since $1 \leq p < \infty$ and $0 < \beta \leq 1$, it follows from this that $k \geq r$ and consequently $f^{(r)}(\lambda) \in MH(k - r, \beta, p, G(a))$. On the basis of the inequality (2.9) we see that

$$\begin{aligned} \inf_{p_n} \| f^{(r)}(\lambda) - p_n(\lambda) \|_{L_p(\Gamma_a)} &= \| f^{(r)}(\lambda) - \pi_n(\lambda; f^{(r)}) \|_{L_p(\Gamma_a)} \\ &\leqslant C(M)\, n^{r-k-\beta} \qquad (n \geqslant k - r), \end{aligned} \tag{3.27}$$

where

$$\| f(\lambda) \|_{L_p(\Gamma_a)} = \left\{ \int_{\Gamma_a} | f(\lambda) |^p \, | d\lambda | \right\}^{\frac{1}{p}}$$

Selecting m so that $2^{m-1} \leq n < 2^m$, and applying the inequality (see for example [13], 1951 ed., p. 357)

$$| p_n(z) | \leqslant C n^{\frac{1}{p}} \| p_n(\lambda) \|_{L_p(\Gamma_a)} \quad (z \in \Gamma_a), \tag{3.28}$$

we obtain

$$\begin{aligned} \Big| \pi_{2^m}(z; f^{(r)}) - \pi_n(z; f^{(r)}) &+ \sum_{l=1}^{\infty} [\pi_{2^{m+l}}(z; f^{(r)}) \\ - \pi_{2^{m+l-1}}(z; f^{(r)})] \Big| &\leqslant C 2^{\frac{m}{p}} \| \pi_{2^m} - f^{(r)} + f^{(r)} - \pi_n \|_{L_p(\Gamma_a)} \\ &+ C \sum_{l=1}^{\infty} 2^{\frac{m+l}{p}} \| \pi_{2^{m+l}} - f^{(r)} + f^{(r)} - \pi_{2^{m+l-1}} \|_{L_p(\Gamma_a)} \leq \end{aligned}$$

[4] It will be obvious from the proof that the integral entering into the definition of $\Psi_a^{(r)}(f)$ is meaningful.

$$\leqslant C(M)\Big\{2^{\frac{m}{p}} n^{r-k-\beta} + \sum_{l=1}^{\infty} 2^{(m+l)\left(\frac{1}{p}+r-k-\beta\right)}\Big\}$$

$$\leqslant C(M)\Big\{2^{\frac{m}{p}} n^{r-k-\beta} + 2^{m\left(\frac{1}{p}+r-k-\beta\right)}\Big\} = O\left(n^{r+\frac{1}{p}-k-\beta}\right) \quad (3.29)$$

when $n > k$ and $z \in \Gamma_a$. Consequently the series in the left-hand side of this inequality converges uniformly when $k+\beta > r+1/p$ in the closed region $\overline{G(a)}$. Since in $L_t(\Gamma_a)$ it converges to $f^{(r)}(\lambda) - \pi_n(\lambda; f^{(r)})$, its sum equals $f^{(r)}(\lambda) - \pi_n(\lambda; f^{(r)})$, and (3.29) may be rewritten in the form

$$|f^{(r)}(\lambda) - \pi_n(\lambda; f^{(r)})| \leqslant C(M)\, n^{r+\frac{1}{p}-k-\beta}. \quad (3.30)$$

Thus if $k+\beta > r+1/p$ the function $f^{(r)}(\lambda)$ is continuous in the region $\overline{G(a)}$. Therefore the formulas and estimates (3.13)—(3.18) obviously remain valid in this case, and if we prove that the component J_1 in equation (3.18) permits the estimate

$$|J_1| \leqslant C(M)\, n^{r+\frac{1}{p}-k-\beta},$$

in the present case, then the theorem will be proved for $a = b$ and $k+\beta > r+1/p$ since we shall have

$$\Psi_a^{(r)}(p_n(f)) = \Phi_a^{(r)}(f) + O\left(n^{r+\frac{1}{p}-k-\beta}\right)$$

in place of formula (3.19).

Let γ_{a_n} be the level line of the domain $G(a)$ for a conformal mapping of it on a unit circle passing through the point $a_n = a - 1/n$. Assuming $q = p/(p-1)$ and considering inequalities (3.27) and (3.30), we obtain

$$|J_1| \leqslant \left| \frac{1}{2\pi i} \int_{\gamma_{a_n}} \frac{f^{(r)}(\lambda) - \pi_n(\lambda; f^{(r)})}{g^{n+1}(\lambda)} \frac{g^{n+1}(a)}{\lambda - a} d\lambda + \pi_n(a; f^{(r)}) - f^{(r)}(a) \right|$$

$$\leqslant C \frac{g^{n+1}(a)}{g^{n+1}(a_n)} \Big\{ \int_{\gamma_{a_n}} |f^{(r)}(\lambda) - \pi_n(\lambda; f^{(r)})|^p d\lambda \Big\}^{\frac{1}{p}} \Big\{ \int_{\gamma_{a_n}} \frac{|d\lambda|}{|\lambda - a|^q} \Big\}^{\frac{1}{q}}$$

$$+ |f^{(r)}(a) - \pi_n(a, f^{(r)})|.$$

$$\leqslant C(a - a_n)^{-\frac{q-1}{q}} \Big\{ \int_{\Gamma_a} |f^{(r)}(\lambda) - \pi_n(\lambda; f^{(r)})|^p |d\lambda| \Big\}^{\frac{1}{p}}$$

$$+ |f^{(r)}(a) - \pi_n(a; f^{(r)})| \leqslant C(M)\, n^{r+\frac{1}{p}-k-\beta}$$

As has already been noted, the statement of the theorem follows from this in case 1).

2) Now let $a = b$ and $k+\beta \leqq r+1/p$. In this case formulas (3.13), (3.14), (3.20) and (3.21) remain valid, but in place of inequality (3.23) we

shall obviously have[5]

$$\int_{-a_0}^{a_0} f^{(r)}(x)\,dh(x) = \int_{-a_0}^{a_0} p_k^{(r)}(x; f)\,dh(x) + O(1). \tag{3.31}$$

Therefore to complete the proof of the theorem it is sufficient to establish that when $k + \beta \leqq r + 1/p$ the sequence $\{F_{r,n}(f)\}$ defined by equation (3.20) admits the estimate

$$F_{r,n}(f) = p_k^{(r)}(a; f) + O\left(n^{r+\frac{1}{p}-k-\beta} + \ln n\right) \tag{3.32}$$

with the usual restrictions on the constant entering into O.

Let the polynomials $\pi_n(\lambda; f^{(k)})$ be selected such that

$$\| f^{(k)}(\lambda) - \pi_n(\lambda; f^{(k)}) \|_{L_p(\Gamma_a)} \leqslant C(M)(n+1)^{-\beta}. \tag{3.33}$$

This may be done since $f^{(k)}(\lambda) \in MH(0, \beta, p, G(a))$, where it is possible to take $p_k^{(k)}(\lambda; f)$ as $\pi_0(\lambda; f^{(k)})$. Assuming the Cauchy formula, we obtain

$$F_{r,n}(f) = \frac{1}{2\pi i} \int_{\gamma_{a_n}} \frac{f^{(k)}(\lambda) - \pi_n(\lambda; f^{(k)})}{g^{n+1}(\lambda)} \frac{d^{r-k}}{da^{r-k}} \left(\frac{g^{n+1}(a)}{\lambda - a} \right) d\lambda + \pi_n^{(r-k)}(a; f^{(k)}) = I_1 + I_2,$$

where γ_{a_n} is the level line of the region $G(a)$ passing through the point $a_n = a - 1/n$. We have

$$|I_1| \leqslant C \frac{g^{n+1}(a)}{g^{n+1}(a_n)} \| f^{(k)}(\lambda) - \pi_n(\lambda; f^{(k)}) \|_{L_p(\Gamma_a)} \sum_{s=0}^{r-k} (n+1)^{r-k-s}$$

$$\times \left\{ \int_{\gamma_{a_n}} \frac{|d\lambda|}{|\lambda - a|^{(s+1)q}} \right\}^{\frac{1}{q}} \leqslant \frac{C(M)}{(n+1)^{\beta}} \sum_{s=0}^{r-k} (n+1)^{r-k-s} (a - a_n)^{-s-1+\frac{1}{q}}$$

$$\leqslant C(M)(n+1)^{r+\frac{1}{p}-k-\beta}$$

Then, assuming that $n_{-1} = 0$, $n_0 = 1$, $n_1 = 2, \cdots, n_l = 2^l, \cdots, n_{m-1} = 2^{m-1} < n_m = n \geqq 2^m$, we obtain

$$I_2 = \pi_n^{(r-k)}(a; f^{(k)}) = p_k^{(r)}(a; f) + \sum_{l=0}^{m} [\pi_{n_l}^{(r-k)}(a; f^{(k)}) - \pi_{n_{l-1}}^{(r-k)}(a; f^{(k)})],$$

from which, applying inequalities (3.22), (3.28) and (3.33), we find that

$$|\pi_n^{(r-k)}(a; f^{(k)}) - p_k^{(r)}(a; f)| \leqslant C \sum_{l=0}^{m} n_l^{r-k} \max_{\lambda \in \Gamma_a} |\pi_{n_l}(\lambda) - \pi_{n_{l-1}}(\lambda)|$$

$$\leqslant C \sum_{l=0}^{m} n_l^{r-k+\frac{1}{p}} \| \pi_{n_l}(\lambda; f^{(k)}) - \pi_{n_{l-1}}(\lambda; f^{(k)}) \|_{L_p(\Gamma_a)} \leqq$$

[5] In the case $r > k$ (3.31) does not differ from (3.23).

$$\leqslant C(M)\sum_{l=0}^{m} 2^{\left(r-k-3+\frac{1}{p}\right)l} \leqslant C(M, r)\left\{2^{m\left(r-k-3+\frac{1}{p}\right)}+m\right\}$$

$$= O\left(n^{r-k-3+\frac{1}{p}}+\ln n\right),$$

where the constant entering into O will depend only on M, k, β, p, r, a. The equality (3.32) follows from this.

THEOREM 3.4. *Let $\{F_n\}$ be a given sequence of positive numbers, let r and k be nonnegative integers, $a>1$, $b>1$, $M>0$, $0<\beta\leqq 1$, $1\leqq p<\infty$, $1\leqq p'\leqq\infty$. If a sequence of polynomials $\{\hat{p}_n(f)\}$ satisfying the inequality*

$$\|f(x)-\hat{p}_n(x;f)\|_{L_{p'}[-1,1]}\leqslant F_n \quad (n=1,2,\ldots),$$

exists for the functions $f(z)\in MH(k,\beta,p,G(b))$ then for the sequence $\{\Psi_a^{(r)}(\hat{p}_n(f))\}$ when $n\to\infty$ we have

$$\Psi_a^{(r)}(\hat{p}_n(f))=\Phi_a^{(r)}(f)$$

$$+O\left\{n^r g^n(a)\left[F_n+\frac{1+\delta_{a\,b}n^{1/p}(1+\delta_{r+\frac{1}{p},\,k+\beta}\ln n)}{n^{k+\beta}g^n(b)}\right]\right\}, \tag{3.34}$$

where $\Phi_a^{(r)}(f)$ has the same value as in Theorem 3.3, and the constant entering into O depends only on the numbers M, k, β, p, b and on the functional $\Psi_a^{(r)}$.

Theorem 3.5 below refers to the case where $f(z)\in MH(0,0,p,G(b))$ $(1\leqq p\leqq\infty)$. It is proved somewhat more simply than the preceding ones, and we shall only state it.

THEOREM 3.5. *Let $\{F_n\}$ be a given sequence of positive numbers, $a>1$, $b>1$, $1\leqq p'\leqq\infty$, $1\leqq p\leqq\infty$, $M>0$ and let $f(z)\in MH(0,0,p,G(b))$. Take*

$$\Phi_a^{(r)}(f)=\begin{cases}\int_{-a}^{a} f^{(r)}(x)\,dh(x), & \text{if } a<b,\\ \Psi_a^{(r)}(p_0(f)), & \text{if } a\geqslant b,\end{cases}$$

where $h(x)$ is the function defining the functional $\Psi_a^{(r)}$. If polynomials $\hat{p}_n(x;f)$ satisfying the inequality

$$\|f(x)-p_n(x;f)\|_{L_{p'}[-1,1]}\leqslant F_n \quad (n=0,1,\ldots),$$

exist for the function $f(z)\in MH(0,0,p,G(b))$ then the following relationships are fulfilled when $n\to\infty$: if $p=\infty$, then

$$\Psi_a^{(r)}(\hat{p}_n(f))=\Phi_a^{(r)}(f)+O\left\{n^r g^n(a)\left[F_n+\frac{1+\delta_{a,\,b}\ln n}{g^n(b)}\right]\right\} \tag{3.35}$$

if $1\leqq p<\infty$, then

$$\Psi_a^{(r)}(\hat{p}_n(f))=\Phi_a^{(r)}(f)+O\left\{n^r g^n(a)\left[F_n+\frac{1+\delta_{a,\,b}n^{\frac{1}{p}}}{g^n(b)}\right]\right\} \tag{3.36}$$

where the constants entering into O depend only on the quantities M, b, p and the functional $\Psi_a^{(r)}$.

In order to conserve space we did not formulate the results individually for the best polynomials $p_n(f,p')$. If we assume $F_n = g^{-n}(b)$ $(n = 0, 1, \cdots)$, then on the basis of Theorem 2.2 it is possible to take the best polynomials $p_n(f,p')$ as the polynomials $\hat{p}_n(f)$ in Theorem 3.5, and we obtain the corresponding formulas for $\Psi_a^{(r)}(p_n(f,p'))$.

Let us note that without those restrictions which we have imposed on the function $h(x)$ generating the functional $\Psi_a^{(r)}$, the results obtained generally speaking are not correct (except for the case noted in the remarks after Definition 3.2). It turns out that the behavior of the sequence

$$\Psi_{a,r}(\hat{p}_n(f)) = \int_{-a}^{a} \hat{p}_n^{(r)}(x; f)\, dh(x) \quad (n = 0, 1, \ldots),$$

along with everything else essentially depends on the properties of the function $h(x)$ in the neighborhood of the points $x = \pm a$. This will be apparent after the following theorem, where we consider the functionals $\Psi_{\alpha,a}^{(r)}$ (see Definition 3.3 and Lemma 3.2).

THEOREM 3.6. *Let $\{F_n\}$ be a given sequence of positive numbers, let r and k be nonnegative integers, $a > 1$, $b > 1$, $\alpha > -1$, $1 \leqq p' \leqq \infty$, $1 \leqq p \leqq \infty$, $M > 0$, and when $k = 0$ let the number β belong to the interval $0 \leqq \beta \leqq 1$, and when $k > 0$, to the interval $0 < \beta \leqq 1$. For each function $f(z) \in MH(k, \beta, p, G(b))$ take*

$$\Phi_{\alpha,a}^{(r)}(f) = \begin{cases} \displaystyle\int_{-a}^{a} f^{(r)}(x)(a^2 - x^2)^{\alpha} \varphi(x)\, dx, & \text{if } a = b \text{ and} \\ & r + \frac{1}{p} - \alpha < k + 1 + \beta \text{ or } a < b; \\ \Psi_{\alpha,a}^{(r)}(p_k(f)), & \text{if } a = b \text{ and} \\ & r + \frac{1}{p} - \alpha \geqslant k + 1 + \beta \text{ or } a > b, \end{cases}$$

where $\varphi(x)$ is the function defining the functional $\Psi_{\alpha,a}^{(r)}$. If polynomials $\hat{p}_n(f)$ such that

$$\| f(x) - \hat{p}_n(x; f) \|_{L_{p'}[-1,1]} \leqslant F_n \quad (n = k, k + 1, \ldots)$$

exist for the function $f(z) \in MH(k, \beta, p, G(b))$, then for $n > k$ the following expressions hold for them: if $p = \infty$, then

$$\Psi_{\alpha,a}^{(r)}(\hat{p}_n(f)) = \Phi_{\alpha,a}^{(r)}(f) + O\left\{n^{r-1-\alpha} g^n(a) \left[F_n + \frac{1 + \delta_{a,b} \ln n}{n^{k+\beta} g^n(b)} \right]\right\}, \tag{3.37}$$

and if $1 \leqq p < \infty$, then

$$\Psi^{(r)}_{\alpha, a}(\hat{p}_n(f)) = \Phi^{(r)}_{\alpha, a}(f)$$
$$+ O\left\{n^{r-1-\alpha}g^n(a)\left[F_n + \frac{1+\delta_{a,b}n^{\frac{1}{p}}(1+\delta_{r+\frac{1}{p}-\alpha, k+1+\beta}\ln n)}{n^{k+\beta}g^n(b)}\right]\right\} \quad (3.38)$$

where the constants entering into O depend only on the numbers M, k, β, b, p and on the functional $\Psi^{(r)}_{\alpha,a}$.

The proof of this theorem when $a \neq b$ does not differ from the proof of Theorem 3.2 except that in place of Lemma 3.1 it is necessary to use Lemma 3.2. Therefore we shall consider only the case $a = b$. Inasmuch as when $a = b$ the theorem is proved basically just as the corresponding results for the functionals $\Psi^{(r)}_a$, we shall not elaborate on the calculations presented.

1) Under the conditions of the theorem let $a = b$ and $r + 1/p - \alpha < k + 1 + \beta$. Let us select the points a_0 and a_1 such that $1 < a_1 < a_0 < a$. The function $(a^2 - x^2)^\alpha\varphi(x)$ will be designated as $\varphi_\alpha(x)$. We have

$$\Psi^{(r)}_{\alpha, a}(p_n) = \int_{-a_0}^{a_0} p^{(r)}_n(x)\varphi_\alpha(x)\,dx + \int_{a_0}^{a} p^{(r)}_n(x)\varphi_\alpha(x)\,d.$$
$$+ \int_{-a}^{-a_0} p^{(r)}_n(x)\varphi_\alpha(x)\,dx = \Psi_{a_0, r}(p_n) + F_r(p_n) + G_r(p_n).$$

Let us first estimate the values of the functional $\Psi^{(r)}_{\alpha,a}$ in the sequence of polynomials $p_n(a; f)$. We have

$$\Psi_{a_0, r}(p_n(f)) = \int_{-a_0}^{a_0} f^{(r)}(x)\varphi_\alpha(x)\,dx + K_1,$$

where

$$|K_1| = \left|\int_{-a_0}^{a_0}\sum_{l=n}^{\infty}[p^{(r)}_{l+1}(x; f) - p^{(r)}_l(x; f)]\,\varphi_\alpha(x)\,dx\right|$$
$$\leqslant C(M, \Psi^{(r)}_{\alpha, a})\sum_{l=n}^{\infty} l^r g^l(a_0)E_l(f)_{C[-1, 1]}$$
$$\leqslant C(M, \Psi^{(r)}_{\alpha, a})\sum_{l=n}^{\infty} l^{r-k-\beta}\frac{g^l(a_0)}{g^l(a)}$$
$$\leqslant C(M, \Psi^{(r)}_{\alpha, a})\,n^{r-k-\beta}\frac{g^n(a_0)}{g^n(a)} = o\left(n^{r-k-\beta+\frac{1}{p}-\alpha-1}\right)$$

Let us now estimate the sequence $F_r(p_n(f))$. When $x > 1$ let $a_x = a - \frac{1}{2}(a - x)$, and let γ_{a_x} be the level line of the domain $G(a)$ with conformal mapping of it on a unit circle. It may be considered that when $x \to a$ the distance $\rho(\Gamma_a, \gamma_{a_x})$ has the same order $(a - x)$ as when $\lambda \in \gamma_{a_x}: |g(\lambda)| \geqq C(a)g(a_x) > 0$. In order not to have to consider the cases $r \leqq k$ and $r > k$

separately when $0 \leqq \beta < 1$ and $r \leqq k+1$, $r > k+1$ when $\beta = 1$, let us assume $s = \max(k+2, r)$. Integrating by parts, we obtain

$$F_r(p_n) = \int_{a_0}^{a} p_n^{(r)}(x)\,\varphi_\alpha(x)\,dx$$

$$= \sum_{l=0}^{s-r-1} p_n^{(r+l)}(a_0)\,v_{l+1}(a_0) + \int_{a_0}^{a} p_n^{(s)}(x)\,v_{s-r}(x)\,dx,$$

where

$$v_0(x) = \varphi_\alpha(x),\ v_l(x) = \int_x^a d\xi_l \int_{\xi_l}^a d\xi_{l-1} \ldots \int_{\xi_2}^a \varphi_\alpha(\xi_1)\,d\xi_1 \quad (l \geqslant 1),$$

and the sum when $r = s$ equals zero. Let the point a_0 be selected so that in the interval $[a_0, a]$ the function $\varphi(x) = \varphi_\alpha(x)\ (a^2 - x^2)^{-\alpha}$ is continuous and consequently bounded. Then when $x \in (a_0, a)$ we have

$$|v_l(x)| \leqslant C(\varphi_\alpha) \int_x^a d\xi_l \ldots \int_{\xi_2}^a (a - \xi_1)^\alpha\,d\xi_1 \leqslant C(\varphi_\alpha)(a-x)^{\alpha+l}.$$

From this we obtain

$$\int_{a_0}^{a} p_n^{(s)}(x; f)\,v_{s-r}(x)\,dx = \int_{a_0}^{a} f^{(s)}(x)\,v_{s-r}(x)\,dx$$

$$- \int_{a_0}^{a} v_{s-r}(x)\,dx \frac{1}{2\pi i} \int_{\Gamma_{a_1}} \frac{p_n^{(s)}(\zeta; f) - f^{(s)}(\zeta)}{g^{n+1}(\zeta)} \frac{g^{n+1}(x)}{\zeta - x}\,d\zeta$$

$$- \int_{a_0}^{a} v_{s-r}(x)\,dx \frac{1}{2\pi i} \int_{\gamma_{a_x}} \frac{f^{(s)}(\lambda)}{g^{n+1}(\lambda)} \frac{g^{n+1}(x)}{\lambda - x}\,d\lambda$$

$$= \int_{a_0}^{a} f^{(s)}(x)\,v_{s-r}(x)\,dx + K_2 + K_3,$$

where [see (3.7)]

$$|K_2| = \left| \int_{a_0}^{a} g^{n+1}(x)\,v_{s-r}(x) \frac{dx}{2\pi} \int_{\Gamma_{a_1}} \sum_{j=n}^{\infty} [p_{j+1}^{(s)}(\zeta; f) - p_j^{(s)}(\zeta; f)]\,g^{-n-1}(\zeta) \frac{d\zeta}{\zeta - x} \right|$$

$$\leqslant \frac{C(M, \Psi_{\alpha,a}^{(r)})}{(a_0 - a_1)\,g^n(a_1)} \sum_{j=n}^{\infty} j^{s-k-\beta} \frac{g^j(a_1)}{g^j(a)} \int_{a_0}^{a} (a-x)^{s-r+\alpha} g^n(x)\,dx$$

$$\leqslant C(M, \Psi_{\alpha,a}^{(r)})\,n^{r-k-\beta-\alpha-1}$$

and according to inequality (2.10)

$$|K_3| \leqslant C(a)\int_{a_0}^{a}\left[\frac{g(x)}{g(a_x)}\right]^n |v_{s-r}(x)|\, \|f^{(s)}(\lambda)\|_{L_p(\gamma_{a_x})}\, \|(\lambda-x)^{-1}\|_{L_q(\gamma_{a_x})}\,dx$$

$$\leqslant C(M, \Psi^{(r)}_{\alpha, a})\int_{a_0}^{a} R^n(x)\,(a-x)^{\alpha+s-r}(a-x)^{\beta-s+k}(a-x)^{\frac{1}{q}-1}\,dx$$

$$= C(M, \Psi^{(r)}_{\alpha, a})\int_{a_0}^{a} R^n(x)\,(a-x)^{\alpha+\beta+k-r-\frac{1}{p}}\,dx \quad \left(\frac{1}{p}+\frac{1}{q}=1,\ 1\leqslant p<\infty\right).$$

Since the function $R(x) = g(x)g^{-1}(a_x)$ satisfies all the conditions of Lemma 3 of [8], and $R(a) = 1$ under the assumption $\alpha + k - r + \beta - 1/p > -1$ (see also (3.7)) when $1 \leqq p < \infty$, we have

$$|K_3| \leqslant C(M, \Psi^{(r)}_{\alpha, a})\, n^{r+\frac{1}{p}-k-\beta-\alpha-1}.$$

If $p = \infty$, then in place of the estimates

$$\|(\lambda-x)^{-1}\|_{L_q(\gamma_{a_x})} \leqslant C(a, q)\,(a-x)^{\frac{1}{q}-1}$$

we have

$$\|(\lambda-x)^{-1}\|_{L_1(\gamma_{a_x})} \leqslant C(a)\,|\ln(a-x)|,$$

which for K_3 gives the estimates

$$|K_3| \leqslant C(M, \Psi^{(r)}_{\alpha, a})\, n^{r-k-\beta-\alpha-1}\ln n \quad (p=\infty).$$

Finally,

$$\left|\sum_{l=0}^{s-r-1} p_n^{(r+l)}(a_0; f)\, v_{l+1}(a_0) - \sum_{l=0}^{s-r-l} f^{(r+l)}(a_0)\, v_{l+1}(a_0)\right|$$

$$\leqslant C(a_0, \Psi^{(r)}_{\alpha, a}) \sum_{l=0}^{s-r-1}\sum_{\nu=n}^{\infty} |p_{\nu+1}^{(r+l)}(a_0; f) - p_\nu^{(r+l)}(a_0; f)|$$

$$\leqslant C(M, \Psi^{(r)}_{\alpha, a}) \sum_{l=0}^{s-r-1}\sum_{\nu=n}^{\infty} (\nu+1)^{r+l} g^{\nu+1}(a_0)\, \nu^{-k-\beta} g^{-\nu}(a)\cdot$$

$$\leqslant C(M, \Psi^{(r)}_{\alpha, a})\, n^{s-1} g^n(a_0)\, g^{-n}(a) = o\left(n^{r+\frac{1}{p}-k-\beta-\alpha-1}\right).$$

From these estimates we find that

$$F_r(p_n(f)) = \sum_{l=0}^{s-r-1} {}^{(l+r)}(a_0)\, v_{l+1}(a_0) + \int_{a_0}^{a} f^{(s)}(x)\, v_{s-r}(x)\,dx$$

$$+\, O\left[n^{r-k-\beta+\frac{1}{p}-\alpha-1}\left(1+\delta_{\frac{1}{p},0}\ln n\right)\right] \qquad (1\leqslant p\leqslant\infty)$$

and, in particular, that the integral in the right-hand side of this equation is finite. Integrating by parts in the reverse order (when $s > r$), we find that

$$F_r(p_n(f)) = \int_{a_0}^{a} f^{(r)}(x)\varphi_\alpha(x)\,dx + O\left\{n^{r+\frac{1}{p}-k-\beta-\alpha-1}\left(1+\delta_{0,\frac{1}{p}}\ln n\right)\right\},$$

where the constant entering into O depends only on the parameters M, a, k, β, p and on the functional $\Psi^{(r)}_{\alpha,a}$.

Analogously it is proved that

$$G_r(p_n(f)) = \int_{-a}^{-a_0} f^{(r)}(x)\varphi_\alpha(x)\,dx + O\left\{n^{r+\frac{1}{p}-k-\beta-\alpha-1}\left(1+\delta_{0,\frac{1}{p}}\ln n\right)\right\}.$$

From this and from the estimates of $\Psi_{a_0,r}(p_n(f))$ and $F_r(p_n(f))$ we find that when $a = b$, $1 \leqq p \leqq \infty$, $k+\beta+\alpha+1-r-1/p > 0$ we have

$$\Psi^{(r)}_{\alpha,a}(p_n(f)) = \int_{-a}^{-a_0} f^{(r)}(x)\varphi_\alpha(x)\,dx + O\left\{n^{r+\frac{1}{p}-k-\beta-\alpha-1}\left(1+\delta_{\frac{1}{p},0}\ln n\right)\right\}$$

and

$$\Psi^{(r)}_{\alpha,a}(\hat{p}_n(f)) = \int_{-a}^{a} f^{(r)}(x)\varphi_\alpha(x)\,dx$$
$$+ O\left\{F_n n^{r-\alpha-1} g^n(a) + n^{r+\frac{1}{p}-k-\beta-\alpha-1}\left(1+\delta_{0,\frac{1}{p}}\ln n\right)\right\}.$$

In case 1) the theorem is proved.

2) Under the conditions of the theorem let $a = b$ and $k+\beta+\alpha+1-r-1/p \leqq 0$. Retaining the former representations of the functionals $\Psi^{(r)}_{\alpha,a}(p_n)$ and $F_r(p_n)$, we estimate the individual components in the following way:

$$\Psi_{a_0,r}(p_n(f)) = \int_{-a_0}^{a_0} p_k^{(r)}(x;f)\varphi_\alpha(x)\,dx$$
$$+ \int_{-a_0}^{a_0}\sum_{\nu=k+1}^{n}[p_\nu^{(r)}(x;f) - p_{\nu-1}^{(r)}(x;f)]\varphi_\alpha(x)\,dx = \int_{-a_0}^{a_0} p_k^{(r)}(x;f)\varphi_\alpha(x)\,d.$$
$$+ O\left(\sum_{\nu=k+1}^{n}\nu^{r-k-\beta}g^\nu(a_0)\,g^{-\nu}(a)\right) = \int_{-a_0}^{a_0} p_k^{(r)}(x;f)\varphi_\alpha(x)\,dx + O(1);$$

analogously

$$\sum_{l=0}^{s-r-1} p_n^{(r+l)}(a_0;f)\,v_l(a_0) = \sum_{l=0}^{r-1} p_k^{(r+l)}(a_0;f)\,v_l(a_0) + O(1).$$

We then assume $\underline{a}_n = a - 3/n$, $a_n = a - 2/n$, $\bar{a}_n = a - 1/n$ $(n > 1)$. When $n > 3(a-a_0)^{-1}$ we have

$$\int_{a_0}^{a} p_n^{(s)}(x; f) v_{s-r}(x)\,dx = -\frac{1}{2\pi i}\int_{a_0}^{a} v_{s-r}(x)\,dx \int_{\Gamma_{a_1}} \frac{p_n^{(s)}(\lambda; f) - f^{(s)}(\lambda)}{\lambda - x}\,\frac{g^{n+1}(x)}{g^{n+1}(\lambda)}\,d\lambda$$

$$-\frac{1}{2\pi i}\left(\int_{a_0}^{a_n} + \int_{a_n}^{a}\right) v_{s-r}(x)\,dx \int_{\Gamma_{a_1}} \frac{f^{(s)}(\lambda)}{\lambda - x}\,\frac{g^{n+1}(x)}{g^{n+1}(\lambda)}\,d\lambda$$

$$= K_2 + \int_{a_0}^{a_n} f^{(s)}(x) v_{s-r}(x)\,dx - \frac{1}{2\pi i}\int_{a_0}^{a_n} v_{s-r}(x)\,dx \int_{\gamma_{\bar{a}_n}} \frac{f^{(s)}(\lambda)}{\lambda - x}\,\frac{g^{n+1}(x)}{g^{n+1}(\lambda)}\,d\lambda$$

$$-\frac{1}{2\pi i}\int_{a_n}^{a} v_{s-r}(x)\,dx \int_{\gamma_{\underline{a}_n}} \frac{f^{(s)}(\lambda)}{\lambda - x}\,\frac{g^{n+1}(x)}{g^{n+1}(\lambda)}\,d\lambda = K_2 + K_4 + K_5 + K_6.$$

The estimate of the component K_2 presented in §1 is also correct in the present case. The components K_5 and K_6 are evaluated in the same way as K_3 in the preceding case:

$$|K_5| \leqslant C \int_{a_0}^{a_n} (a - x)^{\alpha+s-r} \frac{g^{n+1}(x)}{g^{n+1}(\bar{a}_n)} \|(\lambda - x)^{-1}\|_{L_q(\gamma_{\bar{a}_n})}\,dx\,\|f^{(s)}\|_{L_p(\gamma_{\bar{a}_n})}$$

$$\leqslant C\left[(\bar{a}_n - a_n)^{-\frac{1}{p}} + \delta_{0,\frac{1}{p}}|\ln(\bar{a}_n - a_n)|\right](a - \bar{a}_n)^{\beta-s+k} g^{-n}(a)$$

$$\times \int_{a_0}^{a} (a - x)^{\alpha+s-r} g^n(x)\,dx \leqslant C(M, \Psi_{\alpha,a}^{(r)})\, n^{r+\frac{1}{p}-k-\beta-\alpha-1}\left(1 + \delta_{0,\frac{1}{p}} \ln n\right);$$

$$|K_6| \leqslant C \int_{a_n}^{a} (a - x)^{\alpha+s-r} \frac{g^{n+1}(x)}{g^{n+1}(\underline{a}_n)} \|(\lambda - x)^{-1}\|_{L_q(\gamma_{\underline{a}_n})}\,dx\,\|f^{(s)}\|_{L_p(\gamma_{\underline{a}_n})}$$

$$\leqslant C(M, \Psi_{\alpha,a}^{(r)})\, n^{r+\frac{1}{p}-k-\beta-\alpha-1}\left(1 + \delta_{0,\frac{1}{p}} \ln n\right).$$

For the estimate of K_4 we shall use the inequality (2.11) of Theorem 2.4, which in our case implies that if $f(z) \in MH(k, \beta, p, G(a))$ and $s \geqq k + 2$, then

$$|f^{(s)}(x)| \leqslant C(M)(a - x)^{-s-k-\frac{1}{p}}$$

Therefore

$$|K_4| \leqslant C \int_{a_0}^{a_n} (a - x)^{\alpha+s-r} |f^{(s)}(x)|\,dx \leqslant C \int_{a_0}^{a_n} (a - x)^{\alpha+\beta+k-r-\frac{1}{p}}\,dx$$

$$\leqslant C\left\{(a - a_n)^{\alpha+1+\beta+k-r-\frac{1}{p}} + \delta_{\alpha+\beta+k,\, r+\frac{1}{p}-1}|\ln(a - a_n)|\right\}$$

$$\leqslant C(M, \Psi_{\alpha,a}^{(r)})\left\{n^{r+\frac{1}{p}-k-\beta-\alpha-1} + \delta_{\alpha+\beta+k,\, r+\frac{1}{p}-1} \ln n\right\}.$$

The sequence $G_r(p_n(f))$ is estimated in exactly the same way. Combining these estimates, we obtain

$$\Psi^{(r)}_{\alpha, a}(p_n(f)) = \int_{-a_0}^{a_0} p_k^{(r)}(x; f)\,\varphi_\alpha(x)\,dx + \sum_{l=0}^{s-r-1} p_k^{(r+l)}(a_0; f) v_l(a_0)$$

$$+ \sum_{l=0}^{s-r-1} (-1)^l p_k^{(r+l)}(-a_0; f)\, w_{l+1}(a_0)$$

$$+ O\left\{n^{r+\frac{1}{p}-k-\beta-\alpha-1}\left(1 + \delta_{0, \frac{1}{p}} \ln n + \delta_{\alpha+\beta+k,\, r+\frac{1}{p}-1} \ln n\right)\right\},$$

where

$$w_l(x) = \int_{-a}^{x} d\xi_l \int_{-a}^{\xi_l} d\xi_{l-1} \ldots \int_{-a}^{\xi_2} \varphi_\alpha(\xi_1)\, d\xi_1.$$

Since $p_k^{(s)}(x) \equiv 0$, we obtain

$$\Psi^{(r)}_{\alpha, a}(p_n(f)) = \Psi^{(r)}_{\alpha, a}(p_k(f))$$

$$+ O\left\{n^{r+\frac{1}{p}-k-\beta-\alpha-1}\left(1 + \delta_{0, \frac{1}{p}} \ln n + \delta_{\alpha+\beta+k,\, r+\frac{1}{p}-1} \ln n\right)\right\}.$$

Just as in case 1, the remaining statements of the theorem follow when $a = b$ and $r + 1/p - k - \beta - \alpha - 1 \geqq 0$.

Up to now we have considered functionals $\Psi(p_n)$ for which the norm $\|\Psi\|_{n,p}$ increases approximately as a geometric progression. For completeness of the picture we shall prove two theorems where the sequence $\{\|\Psi\|_{n,p}\}$ has exponential growth.

THEOREM 3.7. *Let $\{F_n\}$ be a given sequence of positive numbers, $\nu > 0$, $1 \leqq p' \leqq \infty$, and let the parameters M, k, β, p, b be the same as in Theorem 3.6. Let $\Psi(p_n)$ be any homogeneous additive functional defined in the linear manifold of all algebraic polynomials for which*

$$\|\Psi\|_{n, p'} = O(n^\nu) \qquad (n \to \infty).$$

If for the function $f(z) \in MH(k, \beta, p, G(b))$ there exist polynomials $\hat{p}_n(f)$ such that

$$\|f(x) - \hat{p}_n(x; f)\|_{L_{p'}[-1, 1]} \leqslant F_n \quad (n = k, k+1, \ldots),$$

then for them

$$\Psi(\hat{p}_n(f)) = \Phi(f) + O\{n^\nu[F_n + n^{-k-\beta} g^{-n}(b)]\}, \tag{3.39}$$

where it is assumed that

$$\Phi(f) = \lim_{n \to \infty} \Psi(p_n(f, p')), \tag{3.40}$$

and the constant entering into O does not depend on the function $f(z)$ and the sequence $\{F_n\}$.

As will be obvious from the proof, under the enumerated conditions the limit (3.40) always exists.

PROOF. Let us first consider the case where $F_n = Cn^{-k-\beta}g^{-n}(b)$ $(n = 1, 2, \cdots)$. On the basis of the results of §2, for any function $f \in MH(k, \beta, p, G(b))$ we have

$$\| f(x) - p_n(x; f, p') \|_{L_{p'}[-1,1]} = E_n(f)_{p'} \leqslant C(M)\, n^{-k-\beta} g^{-n}(b) = F_n.$$

Therefore the series

$$\Psi(p_k(f, p')) + \sum_{l=k+1}^{\infty} \Psi(p_l(f, p') - p_{l-1}(f, p'))$$

is majorized by the converging series

$$\sum_{l=k+1}^{\infty} l^{\nu-k-\beta} g^{-l}(b),$$

and consequently the limit (3.40) exists. From this we obtain

$$|\Phi(f) - \Psi(p_n(f, p'))| \leqslant C(\Psi, M) \sum_{l=n+1}^{\infty} l^{\nu-k-\beta} g^{-l}(b) = O(n^{\nu-k-\beta} g^{-n}(b)).$$

Now let $\{F_n\}$ be any sequence $(F_n > 0)$ and let the polynomials $\hat{p}_n(f)$ satisfy the conditions of the theorem. We have

$$|\Phi(f) - \Psi(\hat{p}_n(f))| \leqslant |\Phi(f) - \Psi(p_n(f, p'))| + |\Psi(p_n(f, p')) - \Psi(\hat{p}_n(f))| = O(n^{\nu-k-\beta} g^{-n}(b) + n^{\nu} F_n),$$

and the theorem is proved.

Let $k \geqq 0$ be an integer, $0 < \beta \leqq 1$. Let us denote by $Mh(k, \beta, p)$ the class of functions $f(x) \in L_p[-1, 1]$ for which

$$E_n(f)_p \leqslant M n^{-k-\beta} \qquad (n \geqslant k).$$

Let us estimate the value of $\Psi(\hat{p}_n(f))$ if $f \in Mh(k, \beta, p)$.

THEOREM 3.8. *Let $\{F_n\}$ be a given sequence of positive numbers, let $k \geqq 0$ be an integer, $M > 0$, $0 < \beta \leqq 1$, $1 \leqq p \leqq \infty$, $\nu > 0$, and let $\Psi(p_n)$ be any additive homogeneous functional satisfying the condition*

$$\| \Psi \|_{n, p} = O(n^{\nu}) \qquad (n \to \infty).$$

For any function $f(x) \in Mh(k, \beta, p)$ assume that

$$\Phi(f) = \begin{cases} \lim_{l \to \infty} \Psi(p_{2^l}(f, p)), & \text{if } k + \beta > \nu,^{6} \\ \Psi(p_k(f, p)), & \text{if } k + \beta \leqslant \nu. \end{cases}$$

[6] The fact that this limit exists follows from the proof.

If there exist polynomials $\hat{p}_n(f)$ *such that*

$$\|f(x) - \hat{p}_n(x;f)\|_{L_p[-1,1]} \leqslant F_n \quad (n = k, k+1, \ldots),$$

for the function $f(x) \in Mh(k, \beta, p)$, *then when* $n \to \infty$

$$\Psi(\hat{p}_n(f)) = \Phi(f) + O\left\{n^\nu\left(F_n + \frac{1+\delta_{\nu,k+\beta}\ln n}{n^{k+\beta}}\right)\right\}, \tag{3.41}$$

where the value of O *is uniform with respect to* $f \in Mh(k, \beta, p)$ *and with respect to the sequences* $\{F_n^m\}$.

PROOF. Let $k + \beta \leqq \nu$, and let $m = m(k)$ and $s = s(n)$ be integers such that $2^{m-1} \leqq k < 2^m$, $2^{m+s} < n \leqq 2^{m+s+1}$. Let us then assume that $n_{-1} = k$, $n_0 = 2^m$, $n_1 = 2^{m+1}, \cdots, n_s = 2^{m+s}$, $n_{s+1} = n$. Obviously, we have

$$\begin{aligned} |\Psi(p_n(f,p)) - \Psi(p_k(f,p))| &= \left|\sum_{j=0}^{s+1} \Psi(p_{n_j}(f,p) - p_{n_{j-1}}(f,p))\right| \\ \leqslant 2\sum_{j=0}^{s+1} \|\Psi\|_{n_j,p} E_{n_{j-1}}(f)_p &\leqslant 2^{1-k-\beta} MC(\Psi) \sum_{j=0}^{s+1} 2^{(m+j)(\nu-k-\beta)} \\ &\leqslant \begin{cases} C2^{(m+s)(\nu-k-\beta)} < Cn^{\nu-k-\beta}, & \text{if } \nu > k+\beta, \\ C(s+2) < C\dfrac{\ln n + m\ln 2}{\ln 2} < C\ln n, & \text{if } \nu = k+\beta. \end{cases} \end{aligned} \tag{3.42}$$

Let $k + \beta > \nu$. Selecting the number m such that $2^m \leqq n < 2^{m+1}$, let us assume $n_m = n$, $n_l = 2^l$ $(l > m)$. Let us consider the series

$$\Psi(p_n(f,p)) + \sum_{l=m}^{\infty} \Psi(p_{n_{l+1}}(f,p) - p_{n_l}(f,p)). \tag{3.43}$$

Since $k + \beta > \nu$ this series is majorized by the convergent series $\sum_{l=m}^{\infty} 2^{l(\nu-k-\beta)}$. The partial sums of the series (3.43) have the form $\Psi(p_{2^l}(f,p))$. Therefore it converges, and its sum is $\Phi(f)$. Consequently, when $k + \beta > \nu$ we get

$$\begin{aligned} |\Psi(p_n(f,p)) - \Phi(f)| &= \left|\sum_{l=m}^{\infty} \Psi(p_{n_{l+1}}(f,p) - p_{n_l}(f,p))\right| \\ &\leqslant C\sum_{l=m}^{\infty} 2^{l(\nu-k-\beta)} = C2^{-m(\nu-k-\beta)} \leqslant C(\Psi, M)\, n^{-k-\beta+\nu}. \end{aligned} \tag{3.44}$$

Let the sequence $\{\hat{p}_n(f)\}$ satisfy the conditions of the theorem. From the estimates of (3.42) and (3.44) we obtain

$$\begin{aligned} \Psi(\hat{p}_n(f)) &= \Phi(f) + O(\|\Psi\|_n \|\hat{p}_n(f) - p_n(f,p)\|_{L_p[-1,1]}) \\ &= \Phi(f) + O\left\{n^\nu\left(F_n + \frac{1+\delta_{k+\beta,\nu}\ln n}{n^{k+\beta}}\right)\right\}, \end{aligned}$$

and the theorem is proved.

Just as in all the preceding theorems, the estimate presented in Theorem 3.8 automatically becomes rough if the sequence $\{F_n\}$ vanishes quickly, for example if $F_n = O(n^{-1-k-\beta})$. However, as will follow from the results of §4, in the general case in the class $Mh(k,\beta,p)$ the order of the remainder term in formula (3.41) cannot be reduced when $n \to \infty$. Moreover, when $0<\beta<1$ formula (3.41) will obviously also be exact in the indicated sense in the class of functions $f(x)$ for which $f^{(k)}(x) \in \mathrm{Lip}(\beta,p)$. If $f(x) \in \mathrm{Lip}(1,p)$ formula (3.41), generally speaking, becomes rough. Thus, if

$$\Psi(p_n) = p'_n(x_0) \quad (x_0 \in (-1,1),\ n = 0, 1, \ldots)$$

then from the inequality

$$|p'_n(x)| \leqslant \frac{2}{2\sqrt{1-x^2}}\,\omega\left(2\sin\frac{\pi}{2n};\, p_n\right)_p \qquad (|x|<1)$$

of S. B. Stečkin [22] it follows that

$$\begin{aligned}|p'_n(x_0, f)| = |\Psi(p_n(f))| &\leqslant Cn\omega\left(\frac{1}{n};\, p_n(f) - f + f\right)_p \\ &\leqslant Cn\left[E_n(f)_p + \omega\left(\frac{1}{n};\, f\right)_p\right] = O(1) \qquad (f \in \mathrm{Lip}(1,p)).\end{aligned}$$

And from formula (3.41) it follows only that

$$p'_n(x_0;\ f) = O(\ln n) \quad (f \in \mathrm{Lip}(1,p)).$$

§4. Approximation of functions by algebraic polynomials with constraints

Let Ψ be a nonzero homogeneous additive functional defined on the linear manifold of all algebraic polynomials, let $\rho_n (n = 0, 1, \cdots)$ be the given sequence of real numbers, and let $f(x)$ be a function from $L_p[-1,1]$ $(1 \leqq p \leqq \infty)$. Assume that

$$E_n(f, \Psi, \rho_n)_p = \inf_{\Psi(p_n)=\rho_n} \|f(x) - p_n(x)\|_{L_p[-1,1]},$$

$$E_n(f, \Psi, \rho_n) = E_n(f, \Psi, \rho_n)_\infty.$$

Aside from this, we shall retain the notation of §§2 and 3.

In this section the behavior of the best approximations of $E_n(f,\Psi,\rho_n)_p$ will be investigated for the same classes of functions and the same functionals as were considered in §3. All the formulas for $E_n(f,\Psi,\rho_n)_p$ presented in the theorems of this section (except Theorem 4.2) either are asymptotic or are upper bounds which are exact when $\rho_n \equiv \dot{\rho}$ in the sense of order when $n \to \infty$.

With the help of Lemma 3.3 the following analog of Jackson's Theorem may be obtained.

THEOREM 4.1. *Let $1 \leqq p \leqq \infty$, $a > 1$, let $k \geqq 0$ be an integer and let ν be a real number. Let $\Psi(p_n)$ be any additive homogeneous functional satisfying the condition*

$$\|\Psi\|_{n,p} \asymp n^{\nu} g^{n}(a) \quad (n \to \infty).$$

If the derivative $f^{k-1}(x)$ *is absolutely continuous and* $f^{(k)}(x) \in L_p[-1,1]$, *then for* $n > k$

$$E_n(f, \Psi, \rho_n)_p \leqslant K(k, \Psi)\left\{\frac{1}{\varepsilon^{k+1}}\frac{1}{n^k}\omega\left(\frac{1}{n}; f^{(k)}\right)_p\right.$$
$$\left.+\omega\left(\frac{1}{k+1}; f^{(k)}\right)_p g^{(\varepsilon-1)n}(a) + \frac{|\Psi(p_k(f,p)) - \rho_n|}{n^{\nu} g^n(a)}\right\}, \qquad (4.0)$$

where $K(k,\Psi)$ *does not depend on the function* $f(x)$, *and* ϵ *is any number from the interval* $(0,1)$.

PROOF. Applying estimates (1.7), (2.1), (2.2) and Lemma 3.3, we obtain

$$E_n(f, \Psi, \rho_n)_p \leqslant C(k, \Psi)\left\{\omega\left(\frac{1}{k+1}; f^{(k)}\right)_p g^{(\varepsilon-1)n}(a)\right.$$
$$\left.+([\varepsilon n]+1)^{-k}\omega\left(\frac{1}{[\varepsilon n]+1}; f^{(k)}\right)_p\right\} + \frac{|\Psi(p_k(f)) - \rho_n|}{C(\Psi)\, n^{\nu} g^n(a)},$$

from which we have the inequality (4.0) if we take

$$\omega\left(\frac{1}{[\varepsilon n]-1}; f^{(k)}\right)_p \leqslant \omega\left(\frac{1}{\varepsilon n}; f^{(k)}\right)_p \leqslant \frac{2}{\varepsilon}\omega\left(\frac{1}{n}; f^{(k)}\right)_p.$$

If $\Psi = \Psi_{a,r}$ and $k < r$ the estimate obtained will depend only on the smoothness of the function $f(x)$, and if $k \geqq r$ it will also depend on the value of $f(x)$ via the last term. In the case where the first term in the right-hand side of (4.0) is not the principal term when $\rho_n \equiv \rho$, that is, if $f(x) \equiv p_k(x)$, the estimate of (4.0) coincides with the estimate presented below for analytic functions.

Let the homogeneous functional $\Phi(f)$ be defined in the class of functions $H = H(k,\beta,p,G(b))$ (see §2). We shall denote the set of functions from MH for which $\Phi(f) = \rho$ by $MH_{\rho,\Phi}(k,\beta,p,G(b)) = MH_{\rho,\Phi}$.

Obviously, if the class $MH_{\rho_1,\Phi}$ is not empty, then for all $|\rho| < |\rho_1|$ the classes $MH_{\rho,\Phi}$ also are not empty. Let us designate as $\rho(\Phi, M)$ the greatest of the numbers ρ' for which for all $|\rho| < \rho'$ the classes $MH_{\rho,\Phi}$ are not empty. In particular, $\rho(\Phi, M)$ may assume a value of $+\infty$.

Let us proceed to the investigation of the best approximations of $E_n(f,\Psi,\rho_n)_p$ for analytic functions.

THEOREM 4.2. *Let* $k \geqq 0$ *be an integer, let* $0 \leqq \beta \leqq 1$ *when* $k = 0$, $0 < \beta \leqq 1$ *when* $k > 0$, *let* ν *be a real number,* $a > 1$, $b > 1$, $1 \leqq p$, $p' \leqq \infty$, $M > 0$. *Let* $f(z) \in MH(k,\beta,p,G(b))$, *and let* $\Psi(p_n)$ *be any additive homogeneous functional satisfying the condition*

$$\|\Psi\|_{n,p'} \asymp n^{\nu} g^{n}(a) \quad (n \to \infty).$$

Then, in the notation of Theorem 3.1, when $n \to \infty$ *we have*

$$E_n(f, \Psi, \rho_n)_{p'} = \frac{\Phi(f) - \rho_n}{\|\Psi\|_{n,p'}} + O\{g^{-n}(b)[n^{-k-\beta} + \delta_{a,b} n^{-\nu}(n^{\nu-1-k-\beta} + \delta_{k+2,\nu+1} \ln n)]\}, \tag{4.1}$$

where the constant entering into O *depends only on the quantities* M, k, β, p, b *and on the functional* Ψ.

If the parameters $a \neq b$, ν, k, β *and* ρ_n *are such that formula* (4.1) *is not asymptotic, then for* $E_n(f, \Psi, \rho_n)_{p'}$ *it gives an upper bound which is exact when* $n \to \infty$ *in the sense of order with respect to the class* $MH(k, \beta, p, G(b))$. *If* $a < b$ *and* $\rho_n \equiv \rho$, *then for every* $|\rho| < \rho(\Phi, M)$ *the estimate of the remainder term in formula* (4.1) *is also exact in the sense of order. It is impossible to improve it in the class* $MH_{\rho,\Phi}(k, \beta, p, G(b))$.

PROOF. Formula (4.1) follows from equality (1.6′), Theorem 3.1 and from Theorems 2.1 and 2.2 (when $\beta = 0$).

Since the second term in the right-hand side of formula (4.1) is $O(n^{-k-\beta} g^{-n}(b))$ when $a \neq b$, the first statement concerning the accuracy of formula (4.1) follows from the estimate of

$$E_n(f)_{p'} \leqslant E_n(f, \Psi, \rho_n)_{p'}$$

and from Theorems 2.1 and 2.2.

For the proof of the second statement of accuracy of the estimate (4.1) when $a < b$ let us first consider the case $\rho = 0$. Let $p_r(x)$ be a fixed polynomial for which $\Psi(p_r) = 1$, and let $f \in MH(k, \beta, p, G(b))$ and

$$\max_{z \in G(b)} |p_r^{(k+1)}(z)| = \frac{1}{2} M_1 \left(\int_{\Gamma_\theta} ds \right)^{-\frac{1}{p}}.$$

If $\Phi(f) = \rho_1 > 0$ we shall assume that

$$F(z, f) = -\mu_1 p_r(z) + \mu f(z),$$

where $\mu = M(\rho_1 M_1 + 2M)^{-1}$, $\mu_1 = \mu \rho_1$. Considering the definition of the functional $\Phi(f)$ it is easy to prove that $F(z; f) \in MH_{0,\Phi}(k, \beta, p, G(b))$. It is obvious that when $n > r$

$$E_n(F(x; f))_{p'} = \mu E_n(f)_{p'}.$$

Let $0 < \beta \leqq 1$. Let us assume that

$$f_n(x) = \pm C(M) \sum_{l=0}^{\infty} n^{-l(k+\beta+1)} g^{-nl}(b) T'_{nl}(x) \qquad (n = 2, 3, \ldots).$$

When proving Theorem 2.1 it was demonstrated that the number $C(M)$ may be selected such that $f_n(z) \in MH(k, \beta, p, G(b))$ for all $n = 2, 3, \cdots$. The sign in front of $C(M)$ may be selected so that $\Phi(f_n) \geqq 0$. Let us estimate the upper bound of the sequence $\{\Phi(f_n)\}$. Obviously, we have

$$0 \leqslant \Phi(f_n) \leqslant C(M) \sum_{l=0}^{\infty} n^{-l(k+\beta+1)} g^{-n^l}(b) \, | \Psi(T'_{n^l}) |$$

$$\leqslant C(M, \Psi) \sum_{l=0}^{\infty} n^{l(\nu-k-\beta+1)} g^{n^l}(a) \, g^{-n^l}(b) \leqslant C(M, \Psi).$$

Thus there exists a $C \geqq 0$ such that

$$\mu_n = M\,[\Phi(f_n) M_1 + 2M]^{-1} \geqslant C > 0 \qquad (n = 2, 3, \ldots).$$

From this we obtain

$$E_{n-2}(F(x; f_n), \Psi, 0)_{p'} \geqslant E_{n-2}(F(x; f_n))_{p'}$$
$$= \mu_n E_{n-2}(f_n)_{p'} > C n^{-k-\beta} g^{-n}(b) \qquad (n = 2, 3, \ldots).$$

Since $F(z; f_n) \in MH_{0,\Phi}(k, \beta, p, G(b))$, the statement about accuracy of the estimate of (4.1) when $a < b$ and $\rho = 0$ is proved in the case where $0 < \beta \leqq 1$. In the case where $\beta = 0$ the proof is the same except the functions used when proving Theorem 2.2 must be taken as the functions $f_n(z)$ $(n = 2, 3, \cdots)$.

Let $a < b$ and $0 < \rho < \rho(\Phi, M)$. Let us select any number ρ_1 from the interval $(\rho, \rho(\Phi, M))$. By definition of $\rho(\Phi, M)$ a function $f \in MH(k, \beta, p, G(b))$ can be found for which $\Phi(f) = \rho_1$. If

$$E_n\left(\frac{\rho}{\rho_1} f(x); \Psi, \rho\right)_{p'} = o\,(n^{-k-\beta} g^{-n}(b)) \qquad (n \to \infty),$$

we assume that

$$\varphi_n(z) = \frac{\rho}{\rho_1} f(z) + \left(1 - \frac{\rho}{\rho_1}\right) F(z; f_n) \qquad (n = 2, 3, \ldots),$$

where $F(z; f_n)$ are the functions which we constructed when investigating the case $\rho = 0$. Applying Corollary 3.3, it is easy to prove that

$$\varphi_n(z) \in MH_{\rho, \Phi}(k, \beta, p, G(b)) \quad (n = 2, 3, \ldots).$$

In addition,

$$E_{n-2}(\varphi_n(z), \Psi, \rho) \geqslant \left(1 - \frac{\rho}{\rho_1}\right) E_{n-2}(F(x; f_n))_{p'}$$
$$- \frac{\rho}{\rho_1} E_{n-1}(f)_{p'} > C n^{-k-\beta} g^{-n}(b) \qquad (C > 0,\ n = 2, 3, \ldots),$$

and this is what we wish to obtain. If, on the other hand,

$$E_n\left(\frac{\rho}{\rho_1} f(x); \Psi, \rho\right)_{p'} \neq o\,(n^{-k-\beta} g^{-n}(b)) \qquad (n \to \infty),$$

this indicates that the estimate of (4.1) in the class $MH_{\rho,\Phi}$ cannot be improved in the sense of order. The theorem is proved.

If $a = b$ and formula (4.1) is not asymptotic, the estimate given by this formula is obviously rough. In any case, it may be significantly improved for the functionals $\Psi_a^{(r)}$ and $\Psi_{\alpha,a}^{(r)}$. This is done in the following theorems.

THEOREM 4.3. *Let r and k be nonnegative integers, $M>0$, $a>1$, $b>1$, $0<\beta\leqq 1$, $1<p'\leqq\infty$, and let $f(z)\in MH(k,\beta,\infty,G(b))$. Then, in the notation of Theorem 3.2, for any functional $\Psi_a^{(r)}$ when $n\to\infty$ we have*

$$E_n(f,\Psi_a^{(r)},\rho_n)_{p'}=\frac{|\Phi_a^{(r)}(f)-\rho_n|}{\|\Psi_a^{(r)}\|_{n,p'}}+O\left(\frac{1+\delta_{a,b}\ln n}{n^{k+\beta}g^n(b)}\right), \tag{4.2}$$

where the constant entering into O depends only on the quantities M, k, β, b and on the functional $\Psi_a^{(r)}$.

If the parameters a, b, r, k, β and $\rho_n\equiv\rho$ are such that (4.2) is not asymptotic, then for $E_n(f,\Psi_a^{(r)},\rho)_{p'}$ it gives an upper bound which is exact in the sense of order with respect to the class $MH(k,\beta,\infty,G(b))$. If $a=b$ and $r\leqq k$ or $a<b$, then for every $|\rho|<\rho(\Phi_a^{(r)},M)$ the value of the remainder term in formula (4.2) ($\rho_n\equiv\rho$) is also exact in the sense of order. It cannot be improved in the class $MH_{\rho,\Phi_a^{(r)}}(k,\beta,\infty,G(b))$.

PROOF. Equation (4.2) follows directly from formulas (1.6), (3.6), Theorem 2.1 and Corollary 3.4 if we assume $\hat{p}_n(f)=p_n(f,p')$ in the latter.

On the basis of the inequality

$$E_n(f,\Psi_a^{(r)},\rho)_1\leqslant E_n(f,\Psi_a^{(r)},\rho)_{p'}\quad(1\leqslant p'\leqslant\infty)$$

it is sufficient to prove the statements about accuracy of formula (4.2) only when $p'=1$. When $a\neq b$ the proof is just as in Theorem 4.2, and so we shall consider only the case $a=b$.

Let $a=b$ and let the principal term in the right-hand side of formula (4.2) be the second term. This will occur if $r>k$ or if $r\leqq k$ and

$$\int_{-a}^{a}f^{(r)}(x)\,dh(x)=\rho.$$

Assuming for definiteness that $\delta(a)=h(a)-h(a-0)\neq 0$, we take

$$f_{k,\beta}(z)=\hat{p}_k(z)+\sum_{l=1}^{\infty}\nu^{-l(k+\beta)}Q_{\nu^l}(z),$$

where $\nu>1$ is a natural number, $k\geqq 0$ is an integer, $0<\beta\leqq 1$,

$$Q_n(z)=\sum_{j=1}^{n}\frac{1}{j}\frac{T_{n-j}(z)}{g^{n-j}(a)}+\sum_{j=n+1}^{2n}\frac{1}{n-j}\frac{T_j(z)}{g^j(a)}, \tag{4.3}$$

and the polynomial $\hat{p}_k(z)$ of degree k is selected such that when $r\leqq k$

$$\int_{-a}^{a}f_{k,\beta}^{(r)}(x)\,dh(x)=\rho.$$

In the case where $r>k$ we can set $\hat{p}_k(z)\equiv 0$.

We shall show that $f_{k,\beta}(z) \in MH(k,\beta,\infty,G(a))$ $(0<\beta\leq 1)$ and that for a sequence of indices $N_m \uparrow \infty$ the inequalities

$$E_{N_m}(f_{k,\beta}, \Psi_a^{(r)}, \rho) \geqslant CN_m^{-k-\beta} \ln N_m g^{-\nu^m}(a) \quad (C>0), \tag{4.4}$$

are valid. Then Theorem 4.3 will be completely proved.

We have

$$\max_{z\in\overline{G(a)}} |Q_n(z)| \leqslant \frac{1}{2} \max_{\lambda\in\Gamma_a} \left| \sum_{j=1}^{n} \frac{1}{j} \frac{g^{n-j}(\lambda)}{g^{n-j}(a)} + \sum_{j=n+1}^{2n} \frac{1}{n-j} \frac{g^{j}(\lambda)}{g^{j}(a)} \right|$$
$$+\frac{1}{2}\sum_{j=1}^{n}\frac{1}{j}\frac{1}{g^{2(n-j)}(a)}+\frac{1}{2}\sum_{j=n+1}^{2n}\frac{1}{n-j}\frac{1}{g^{2j}(a)}.$$

Assuming $\theta = \arg g(\lambda)$ $(\lambda \in \Gamma_a)$, we obtain from this

$$\max_{z\in G(a)} |Q_n(z)| \leqslant \frac{1}{2} \max_{0\leqslant\theta\leqslant 2\pi} \left| \sum_{j=1}^{n} \frac{e^{i(n-j)\theta}}{j} \right.$$
$$\left. + \sum_{j=n+1}^{2n} \frac{e^{ij\theta}}{n-j} \right| + C \leqslant \max_{0\leqslant\theta\leqslant 2\pi} \left| e^{i\theta n} \sum_{j=1}^{n} \frac{\sin j\theta}{j} \right| + C \leqslant C,$$

since the sequence of polynomials

$$\sum_{j=1}^{n} \frac{\sin j\theta}{j}$$

is uniformly bounded (see for example [23], Chapter I, §30). Therefore, on the basis of (3.22), the series

$$\sum_{l=1}^{\infty} \nu^{-l(k+\beta)} \frac{d^k}{dz^k} Q_{\nu^l}(z)$$

in the domain $\overline{G(a)}$ is majorized by the convergent numerical series

$$\sum_{l=1}^{\infty} \nu^{-l\beta}.$$

Consequently the function $f_{k,\beta}(z)$ is k times continuously differentiable in the closed domain $\overline{G(a)}$. Then let $z_1, z_2 \in \Gamma_a$, $0<\beta<1$ and $1/\nu^{m+1} \leq |z_1-z_2| < 1/\nu^m$.

Using a well-known procedure (see for example [12], 1947 ed., p. 180), we obtain

$$|f_{k,\beta}^{(k)}(z_1) - f_{k,\beta}^{(k)}(z_2)| \leqslant 4|z_1-z_2| \sum_{l=1}^{m} \nu^{-l(k+\beta)} \max_{z\in\Gamma_a} \left| \frac{d^{k+1}}{dz^{k+1}} Q_{\nu^l}(z) \right|$$
$$+ 2\sum_{l=m+1}^{\infty} \nu^{-l(k+\beta)} \max_{z\in\Gamma_a} \left| \frac{d^k}{dz^k} Q_{\nu^l}(z) \right| \leqslant 4C|z_1-z_2| \sum_{l=1}^{m} \nu^{l(1-\beta)} + 2C \sum_{l=m+1}^{\infty} \nu^{-l\beta}$$
$$\leqslant C(a,k,\beta)\left\{|z_1-z_2|\nu^{m(1-\beta)} + \frac{1}{\nu^{(m+1)\beta}}\right\} \leqslant M|z_1-z_2|^{\beta},$$

that is, $f_{k,\beta}(z) \in MH(k, \beta, \infty, G(a))$ when $0 < \beta < 1$.

Let $\beta = 1$ and let s be the length of the arc on the ellipse Γ_a reckoned from some point. Then for any function $\varphi(z)$ analytic in the closed domain $G(a)$ we have

$$|\varphi(s+h) - 2\varphi(s) + \varphi(s-h)| \leqslant Ch^2 \{ \max_{z \in G(a)} |\varphi''(z)| + \max_{z \in G(a)} |\varphi'(z)| \},$$

where C is an absolute constant for fixed a. In fact, since in the equation $z = z(s)$ for the ellipse Γ_a the function $z(s)$ is an analytic function of the parameter s, we see that $\varphi(s)$ also is analytic and

$$|\varphi''(s)| = |z'^2(s)\varphi''(z) + z''(s)\varphi'(z)| \leqslant C(a) \{ \max_{z \in G(a)} |\varphi''(z)| + \max_{z \in G(a)} |\varphi'(z)| \}.$$

Then, applying the Taylor formula separately to the imaginary and real parts of the function $\varphi(s)$, we obtain

$$|\varphi(s+h) - 2\varphi(s) + \varphi(s-h)| \leqslant Ch^2 \max |\varphi''(s)| \leqslant C(a) h^2 \{ \max_{z \in G(a)} |\varphi''(z)| + \max_{z \in G(a)} |\varphi'(z)| \}.$$

Therefore, selecting m such that $\nu^{-(m+1)} < h < \nu^{-m}$, we obtain

$$\begin{aligned} &|f_{k,1}^{(k)}(s+h) - 2f_{k,1}^{(k)}(s) + f_{k,1}^{(k)}(s-h)| \\ &\leqslant Ch^2 \sum_{l=1}^{m} \nu^{-l(k+1)} \left\{ \max_{z \in \Gamma_a} \left| \frac{d^{k+z}}{dz^{k+2}} Q_{\nu l}(z) \right| + \max_{z \in \Gamma_a} \left| \frac{d^{k+1}}{dz^{k+1}} Q_{\nu l}(z) \right| \right\} \\ &+ 4 \sum_{l=m+1}^{\infty} \nu^{-l(k+1)} \max_{z \in \Gamma_a} \left| \frac{d^k}{dz^k} Q_{\nu l}(z) \right| \leqslant C \left\{ h^2 \sum_{l=1}^{m} \nu^l + \sum_{l=m+1}^{\infty} \nu^{-l} \right\} \\ &\leqslant C \{ h^2 \nu^m + \nu^{-(m+1)} \} \leqslant M |h|. \end{aligned}$$

Thus for any $0 < \beta \leqq 1$ there exists an $M > 0$ such that $f_{k,\beta}(z) \in MH(k, \beta, \infty, G(a))$.

We shall assume that $\nu = 2$ in the remainder of this article. Let $N_m = 2^m - 1$. We shall evaluate the lower bound of $E_{N_m}(f_{k,\beta}, \Psi_a^{(r)}, \rho)_1$. Let us assume for this purpose that

$$\begin{aligned} \hat{p}_{N_m}(z) &= \hat{p}_k(z) + \sum_{l=1}^{m} 2^{-l(k+\beta)} \sum_{j=1}^{2^l} \frac{1}{j} \left\{ \frac{T_{2^l - j}(z)}{g^{2^l - j}(a)} - \frac{T_{2^l + j}(z)}{g^{2^l + j}(a)} \right\} \\ &+ \sum_{l=m+1}^{\infty} 2^{-l(k+\beta)} \sum_{j=0}^{2^m - 1} \frac{1}{2^l - j} \frac{T_j(z)}{g^j(a)} + 2^{-(m-1)(k+1+\beta)} \frac{T_{2^m}(z)}{g^{2^m}(a)} \\ &+ 2^{-m(k+\beta)} \sum_{j=1}^{2^m} \frac{1}{j} \frac{T_{2^m + j}(z)}{g^{2^m + j}(a)}. \end{aligned}$$

$\hat{p}_{N_m}(z)$ is a binomial of degree N_m. On the basis of the bounds of (1.8) and (3.6), to prove inequality (4.4) it is sufficient to establish that

$$|\Psi_a^{(r)}(\hat{p}_{N_m})-\rho|>CN_m^{r-k-\beta}\ln N_m \quad (C>0) \tag{4.5}$$

and

$$\|f_{k,\beta}(x)-\hat{p}_{N_m}(x)\|_{L[-1,1]}=o(N_m^{-k-\beta}\ln N_m g^{-N_m}(a)). \tag{4.6}$$

We have

$$\begin{aligned} f_{k,\beta}(z)-\hat{p}_{N_m}(z)= &\sum_{l=m+1}^{\infty}2^{-l(k+\beta)}\sum_{j=2^m}^{2^l-1}\frac{1}{2^l-j}\frac{T_j(z)}{g^j(a)} \\ &-\sum_{l=m}^{\infty}2^{-l(k+\beta)}\sum_{j=1}^{2^l}\frac{1}{j}\frac{T_{2^l+j}(z)}{g^{2^l+j}(a)}-2^{-(m-1)(k+1+\beta)}\frac{T_{2^m}(z)}{g^{2^m}(a)}. \end{aligned} \tag{4.7}$$

Therefore

$$\begin{aligned} \|f_{k,\beta}(x)-\hat{p}_{N_m}(x)\|_{L[-1,1]}\leqslant &\sum_{l=m+1}^{\infty}2^{-l(k+\beta)}\sum_{j=2^m}^{2^l-1}\frac{1}{2^l-j}g^{-j}(a) \\ &+\sum_{l=m}^{\infty}2^{-l(k+\beta)}\sum_{j=1}^{2^l}\frac{1}{j}g^{-(2^l+j)}(a)+2^{-(m-1)(k+1+\beta)}g^{-2^m}(a) \\ &\leqslant C2^{-m(k+\beta)}g^{-2^m}(a)\leqslant CN_m^{k+\beta}g^{-N_m}(a), \end{aligned}$$

and formula (4.6) is proved.

Let us prove inequality (4.5). Take $r>k$. Then

$$\begin{aligned} \Psi_a^{(r)}(\hat{p}_{N_m})-\rho=&\,\delta(a)[\hat{p}_{N_m}^{(r)}(a)-\hat{p}_k^{(r)}(a)] \\ &+\delta(-a)[\hat{p}_{N_m}^{(r)}(-a)-\hat{p}_k^{(r)}(-a)]+O(1)\quad(m\to\infty), \end{aligned}$$

where $\delta(\pm a)$ are the steps of the function $h(x)$ at the points $x=\pm a$. Since

$$T_n^{(r)}(a)=(a^2-1)^{-\frac{r}{2}}n^r g^n(a)\left(1+O\left(\frac{1}{n}\right)\right), \tag{4.8}$$

it is easy to calculate that

$$\begin{aligned} (a^2-1)^{r/2}(\hat{p}_{N_m}^{(r)}(a)-\hat{p}_k^{(r)}(a))=&\sum_{l=1}^{m}2^{-l(k+\beta)}\sum_{j=1}^{2^l}\frac{1}{j}\{(2^l-j)^r \\ -(2^l+j)^r\}+\sum_{l=m+1}^{\infty}2^{-l(k+\beta)}&\sum_{j=0}^{2^m-1}\frac{j^r}{2^l-j}+2^{-m(k+\beta)}\sum_{j=1}^{2^m}\frac{1}{j}(2^m+j)^r \\ &+o(2^{m(r-k-\beta)})=u_1+u_2+u_3+o(N_m^{r-k-\beta})\quad(m\to\infty). \end{aligned}$$

Furthermore,

$$|u_2| \leqslant \sum_{l=m+1}^{\infty} 2^{mr} 2^{-l(k+\beta)} \sum_{j=0}^{2^m-1} \frac{1}{2^l - j} \leqslant 2^{mr} \sum_{l=m+1}^{\infty} 2^{-l(k+\beta)} \left(C + \ln \frac{2^l}{2^l - 2^m} \right)$$
$$= O(N_m^{r-k-\beta}),$$

$$u_1 = -\sum_{l=1}^{m} 2^{-l(k+\beta)} \sum_{j=1}^{2^l} \frac{1}{j} \sum_{s=0}^{r} C_r^s 2^{l(r-s)} [j^s - (-j)^s]$$
$$= -\sum_{s=0}^{\left[\frac{r-1}{2}\right]} 2C_r^{2s+1} \sum_{l=1}^{m} 2^{l(r-k-\beta-2s-1)} \sum_{j=1}^{2^l} j^{2s}.$$

From this, if $r > k + \beta$, then

$$|u_1| \leqslant C \sum_{l=1}^{m} 2^{l(r-k-\beta)} = O(N_m^{r-k-\beta}) \quad (m \to \infty).$$

If $r = k + \beta$ (i.e. $r = k + 1$, $\beta = 1$), then

$$u_1 = -\sum_{s=0}^{\left[\frac{r-1}{2}\right]} 2C_r^{2s+1} \sum_{l=[\sqrt{m}]}^{m} 2^{-l(2s+1)} \sum_{j=1}^{2^l} j^{2s} + O(\sqrt{m})$$
$$= -\sum_{s=0}^{\left[\frac{r-1}{2}\right]} 2C_r^{s+1} \sum_{l=[\sqrt{m}]}^{m} 2^{-l(2s+1)} \left[\frac{2^{l(2s+1)}}{2s+1} + O(2^{2ls}) \right]$$
$$+ O(\sqrt{m}) = -m \sum_{s=0}^{\left[\frac{r-1}{2}\right]} \frac{2C_r^{2s+1}}{2s+1} + O(\sqrt{m})$$
$$= -\frac{2^{r+1}}{r+1} m + O(\sqrt{m}) \quad (m \to \infty).$$

Finally,

$$u_3 = 2^{m(r-k-\beta)} \sum_{s=0}^{r} C_r^s \sum_{j=1}^{2^m} 2^{-ms} j^{s-1} = 2^{m(r-k-\beta)} \sum_{j=1}^{2^m} \frac{1}{j}$$
$$+ O(2^{m(r-k-\beta)}) = \ln 2 \cdot 2^{m(r-k-\beta)} m + O(N_m^{r-k-\beta}).$$

In conclusion we find that for $r > k$

$$(a^2 - 1)^{r/2} (\hat{p}_{N_m}^{(r)}(a) - \hat{p}_k^{(r)}(a)) = \begin{cases} 2^{m(r-k-\beta)} \ln 2^m + O(N_m^{r-k-\beta}), & r > k + \beta \\ -\left[\frac{2^{r+1}}{r+1} - \ln 2 \right] m + O(\sqrt{m}), & r = k + \beta. \end{cases}$$

Considering that $g(-a) = -g(a)$ and that consequently for the analogous bound of the sequence $\hat{p}_{N_m}^{(r)}(-a) - \hat{p}_k^{(r)}(-a)$ sign-variable terms are obtained (when summing with respect to j), we easily find that

$$\hat{p}^{(r)}_{N_m}(-a)-\hat{p}^{(r)}_{k}(-a)=O(N_m^{r-k-\beta}).$$

Therefore, since $\delta(a)\neq 0$, inequality (4.5) may be considered proved when $r>k$.

Let $r\leqq k$. In this case, by assumption,

$$\rho=\int_{-a}^{a} f^{(r)}_{k,\beta}(x)\,dh(x)$$

and consequently

$$\Psi^{(r)}_a(\hat{p}_{N_m})-\rho=\int_{-a}^{a}[\hat{p}^{(r)}_{N_m}(x)-f^{(r)}_{k,\beta}(x)]\,dh(x).$$

From this and from (4.7), by analogy with the preceding, we obtain

$$\begin{aligned}\Psi^{(r)}_a(\hat{p}_{N_m})-\rho=&-\frac{\delta(a)}{(a^2-1)^{r/2}}\Big\{\sum_{l=m+1}^{\infty}\frac{1}{2^{l(k+\beta)}}\sum_{j=1}^{2^l-2^m}\frac{1}{j}[(2^l-j)\\&-(2^l+j)^r]-\sum_{l=m+1}^{\infty}\frac{1}{2^{l(k+\beta)}}\sum_{j=2^l-2^m+1}^{2^l}\frac{1}{j}(2^l+j)^r-\sum_{j=1}^{2^m}\frac{(2^m+j)^r}{j}\frac{1}{2^{m(k+\beta)}}\Big\}\\&+\frac{\delta(-a)}{(a^2-1)^{r/2}}\Big\{-\sum_{l=m+1}^{\infty}\frac{1}{2^{l(k+\beta)}}\sum_{j=1}^{2^l-2^m}\frac{(-1)^j}{j}[(2^l-j)^r-(2^l+j)^r]\\&+\sum_{l=m+1}^{\infty}\frac{1}{2^{l(k+\beta)}}\sum_{j=2^l-2^m+1}^{2^l}\frac{(-1)^j}{j}(2^l+j)^r+\frac{1}{2^{m(k+\beta)}}\sum_{j=1}^{2^m}\frac{(-1)^j}{j}(2^m+j)^r\Big\}\\&+O(2^{m(r-k-\beta)})=\frac{\delta(a)}{(a^2-1)^{r/2}}2^{m(r-k-\beta)}\sum_{j=1}^{2^m}\frac{1}{j}+O(2^{m(r-k-\beta)})\\&=\frac{\delta(a)}{(a^2-1)^{r/2}}N_m^{r-k-\beta}\ln N_m+O(N_m^{r-k-\beta}).\end{aligned}$$

Thus the inequality (4.5) and the theorem are completely proved.

The following two theorems refer to the case where $f(z)\in MH(k,\beta,p,G(b))$ $(1\leqq p<\infty)$ and as before $\Psi=\Psi^{(r)}_a$.

Theorem 4.4. *Let r and k be nonnegative integers, $M>0$, $a>1$, $b>1$, $0<\beta\leqq 1$, $1\leqq p<\infty$, $1\leqq p'\leqq\infty$, and let $f\in MH(k,\beta,p,G(b))$. Then, in the notation of Theorem 3.3, for any functional $\Psi^{(r)}_a$ when $n\to\infty$ we have*

$$E_n(f,\Psi^{(r)}_a,\rho_n)_{p'}=\frac{|\Phi^{(r)}_a(f)-\rho_n|}{\|\Psi^{(r)}_a\|_{n,p'}}+O\left\{\frac{1+\delta_{a,b}n^{1/p}(1+\delta_{r+\frac{1}{p},\,k+\beta}\ln n)}{n^{k+\beta}g^n(b)}\right\},\tag{4.9}$$

where the constant entering into O depends only on the quantities M, k, β, p, b and on the functional $\Psi^{(r)}_a$.

If the parameters a, b, r, k, β, p and $p_n \equiv \rho$ are such that the formula (4.9) is not asymptotic, then for $E_n(f, \Psi_a^{(r)}, \rho)_{p'}$ it gives an upper bound which is exact in the sense of order with respect to the class $MH(k, \beta, p, G(b))$. If $a = b$ and $r + 1/p < k + \beta$, or if $a < b$, then for every $|\rho| < \rho(\Phi_a^{(r)}, M)$ the value of the remainder term in formula (4.9) is also exact in the sense of order. It cannot be improved in the class $MH_{\rho, \Phi_a^{(r)}}(k, \beta, p, G(b))$.

PROOF. Formula (4.9) follows from equality (1.6′), Theorem 2.1, formula 3.6 and Theorem 3.4. Just as in Theorem 4.3 it is sufficient to prove the statements about accuracy of formula (4.9) in the case where $a = b$, $p' = 1$.

Let $a = b$, $1 \leq p < \infty$, $1/p + 1/q = 1$ and, for definiteness, $\delta(a) = h(a) - h(a - 0) \neq 0$. Assume that

$$K_n(z) = n^{-\frac{1}{q}} \sum_{s=1}^{n-1} \left(1 - \frac{s}{n}\right) \left[\frac{T_{n-s}(z)}{g^{n-s}(a)} + \frac{T_{n+s}(z)}{g^{n+s}(a)}\right],$$

$$f_{k,\beta}(z) = \overline{p}_k(z) + \sum_{l=1}^{\infty} 2^{-l(k+\beta)} K_{2^l}(z) \equiv f(z; k, \beta, p, a). \tag{4.10}$$

Since when $z \in \Gamma_a$ we have

$$\frac{1}{2} n^{1/q} e^{-in\theta} K_n(z) = \sum_{s=1}^{n-1} \left(1 - \frac{s}{n}\right) \cos s\theta + O(1) \quad (n \to \infty),$$

where $\theta = \arg g(z)$, and the sum, as we shall see, is the Fejér kernel without a free term, it follows that

$$\max_{z \in \Gamma_a} |K_n(z)| = O\left(n^{\frac{-1}{p}}\right)$$

and

$$\|K_n(z)\|_{L_p(\Gamma_a)} = O\left(n^{\frac{-1}{q}}\right) \left\{\int_0^{2\pi} \left|\frac{\sin^2 n \frac{\theta}{2}}{n \sin^2 \frac{\theta}{2}}\right|^p d\theta\right\}^{\frac{1}{p}} = O(1,$$

Using the same arguments as in Theorem 4.3, we can obtain from this that $f_{k,\beta}(z) \in MH(k, \beta, p, G(a))$ $(0 < \beta \leq 1)$ and that when $r + 1/p < k + \beta$ the function $f_{k,\beta}(z)$ is r times continuously differentiable in the closed domain $\overline{G(a)}$, and consequently $\overline{p}_k(z)$ may be selected so that $\Phi_a^{(r)}(f_{k,\beta}) = \rho$.

Let $N_m = 2^m - 1$, and

$$\overline{p}_{N_m}(z) = \overline{p}_k(z) + \sum_{l=1}^{m-1} 2^{-l\left(k+\beta+\frac{1}{q}\right)} \sum_{j=1}^{2^l-1} \left(1 - \frac{j}{2^l}\right) \left\{\frac{T_{2^l-j}(z)}{g^{2^l-j}(a)} + \frac{T_{2^l+j}(z)}{g^{2^l+j}(a)}\right\}$$

$$+ \sum_{l=m}^{\infty} 2^{-l\left(k+\beta+\frac{1}{q}\right)} \sum_{j=2^l-2^m+1}^{2^l-1} \left(1 - \frac{j}{2^l}\right) \frac{T_{2^l-j}(z)}{g^{2^l-j}(a)}.$$

We have

$$\|f_{k,\beta}(x) - \bar{p}_{N_m}(x)\|_{L[-1,1]} \leqslant \Big\| \sum_{l=m+1}^{\infty} 2^{-l\left(k+\beta+\frac{1}{q}\right)} \sum_{j=1}^{2^l-2^m} \left(1-\frac{j}{2^l}\right)\left\{\frac{T_{2^l-j}(z)}{g^{2^l-j}(a)} + \frac{T_{2^l+j}(z)}{g^{2^l+j}(a)}\right\} + \sum_{l=m}^{\infty} 2^{-l\left(k+\beta+\frac{1}{q}\right)} \sum_{j=2^l-2^m+1}^{2^l-1} \left(1-\frac{j}{2^l}\right)\frac{T_{2^l+j}(z)}{g^{2^l+j}(a)}\Big\|_{C[-1,1]} \tag{4.11}$$

$$\leqslant C g^{-2^m}(a) \sum_{l=m}^{\infty} 2^{-l\left(k+\beta+\frac{1}{q}\right)} = O\left(g^{-N_m}(a)\, N_m^{\frac{1}{p}-k-\beta-1}\right).$$

Let us now evaluate the lower bound of the sequence $\{|\Psi_a^{(r)}(\bar{p}_{N_m}) - \rho|\}$ considering $\rho = \rho_n$ independent of n. If $r + 1/p \geqq k + \beta$, then by analogy with the proof of Theorem 4.3, considering formula (4.8), we obtain

$$\Psi_a^{(r)}(\bar{p}_{N_m}) = \int_{-a_0}^{a_0} \bar{p}_{N_m}^{(r)}(x)\, dh(x) + \delta(a)\, \bar{p}_{N_m}^{(r)}(a)$$

$$+ \delta(-a)\, \bar{p}_{N_m}^{(r)}(-a) = O(1)\left\{\sum_{l=1}^{m-1} 2^{-l\left(k+\beta+\frac{1}{q}\right)} \sum_{j=1}^{2^l-1}\left[(2^l-j)^r\left(\frac{g(a_0)}{g(a)}\right)^{2^l-j}\right.\right.$$

$$\left.\left. + (2^l+j)^r\left(\frac{g(a_0)}{g(a)}\right)^{2^l+j}\right] + \sum_{l=m}^{\infty} 2^{-l\left(k+\beta+\frac{1}{q}\right)} \sum_{j=2^l-2^m+1}^{2^l-1} (2^l-1)^r\left(\frac{g(a_0)}{g(a)}\right)^{2^l-j}\right\}$$

$$+ \rho + O(1)\left\{\sum_{l=1}^{m-1} 2^{-l\left(k+\beta+\frac{1}{q}\right)} \sum_{j=1}^{2^l-1}\left(1-\frac{j}{2^l}\right)[(2^l-j)^{r-1} + (2^l+j)^{r-1}]\right.$$

$$\left. + \sum_{l=m}^{\infty} 2^{-l\left(k+\beta+\frac{1}{q}\right)} \sum_{j=2^l-2^m+1}^{2^l-1}\left(1-\frac{j}{2^l}\right)(2^l-j)^{r-1}\right\}$$

$$+ \frac{\delta(-a)}{(a^2-1)^{r/2}}\left\{\sum_{l=1}^{m-1} 2^{-l\left(k+\beta+\frac{1}{q}\right)} \sum_{j=1}^{2^l-1}(-1)^j\left(1-\frac{j}{2^l}\right)[(2^l-j)^r + (2^l+j)]\right.$$

$$\left. + \sum_{l=m}^{\infty} 2^{-l\left(k+\beta+\frac{1}{q}\right)} \sum_{j=2^l-2^m+1}^{2^l-1}(-1)^j\left(1-\frac{j}{2^l}\right)(2^l-j)^r\right\}$$

$$+ \frac{\delta(a)}{(a^2-1)^{r/2}}\left\{\sum_{l=1}^{m-1} 2^{-l\left(k+\beta+\frac{1}{q}\right)} \sum_{j=1}^{2^l-1}\left(1-\frac{j}{2^l}\right)[(2^l-j)^r + (2^l+j)^r]\right.$$

$$\left. + \sum_{l=m}^{\infty} 2^{-l\left(k+\beta+\frac{1}{q}\right)} \sum_{j=2^l-2^m+1}^{2^l-1}\left(1-\frac{j}{2^l}\right)(2^l-j)^r\right\}.$$

Hence

$$|\Psi_a^{(r)}(\bar{p}_{N_m})-\rho| \geqslant \frac{C\,|\,\delta(a)\,|}{(a^2-1)^{r/2}}\Big\{\sum_{l=1}^{m-1} 2^{l\left(r+1-\frac{1}{q}-k-\beta\right)}$$
$$+\sum_{l=m}^{\infty} 2^{m(r+2)}\, 2^{-l\left(k+\beta+\frac{1}{q}+1\right)}\Big\} + O\big(2^{m\left(r+\frac{1}{p}-1-k-\beta\right)}\big)$$
$$> C\{2^{m\left(r+\frac{1}{p}-k-\beta\right)}+m\} > C\,(N_m^{r+\frac{1}{p}-k-\beta}+\ln N_m), \tag{4.12}$$

where $C>0$. If $r+1/p<k+\beta$, then analogously

$$|\Psi_a^{(r)}(\bar{p}_{N_m})-\rho| = \Big|\int_{-a}^{a}[f_{k,\beta}^{(r)}(x)-\bar{p}_{N_m}^{(r)}(x)]\,dh(x)$$
$$> C\Big\{\sum_{l=m+1}^{\infty} 2^{-l\left(k+\beta+\frac{1}{q}\right)}\sum_{j=1}^{2^l-2^m}\Big(1-\frac{j}{2^l}\Big)[(2^l-j)^r+(2^l+j)^r] \tag{4.13}$$
$$+\sum_{l=m}^{\infty} 2^{-l\left(k+\beta+\frac{1}{q}\right)}\sum_{j=2^l-2^m+1}^{2^l-1}\Big(1-\frac{j}{2^l}\Big)(2^l+j)^r\Big\} > CN_m^{r+\frac{1}{p}-k-\beta}.$$

From (1.8) and (4.11)—(4.13) it follows that

$$E_{N_m}(f_{k,\beta}, \Psi_a^{(r)}, \rho)_1 \geqslant Cg^{-N_m}(a)\frac{1+\delta_{r+\frac{1}{p},\,k+\beta}\ln N_m}{N^{k+\beta-\frac{1}{p}}}, \quad C>0.$$

Theorem 4.4 is proved.

THEOREM 4.5. *Let $a>1$, $b>1$, $M>0$, $1\leqq p\leqq\infty$, $1\leqq p'\leqq\infty$ and let $f(z)\in MH(0,0,p,G(b))$. Then, in the notation of Theorem 3.5, for any functional $\Psi_a^{(r)}$ when $n\to\infty$ we have: if $p=\infty$, then*

$$E_n(f, \Psi_a^{(r)}, \rho_n)_{p'} = \frac{|\Phi_a^{(r)}(f)-\rho_n|}{\|\Psi_a^{(r)}\|_{n,\,p'}} + O\Big(\frac{1+\delta_{a,\,b}\ln n}{g^n(b)}\Big); \tag{4.14}$$

if $1\leqq p<\infty$, then

$$E_n(f, \Psi_a^{(r)}, \rho_n)_{p'} = \frac{|\Phi_a^{(r)}(f)-\rho_n|}{\|\Psi_a^{(r)}\|_{n,\,p'}} + O\Big(\frac{1+\delta_{a,\,b}\,n^{1/p}}{g^n(b)}\Big), \tag{4.15}$$

where the constants entering into O depend only on the quantities M, b, p and on the functional $\Psi_a^{(r)}$.

If the parameters a, b, r, p and $\rho_n\equiv\rho$ are such that these formulas are not asymptotic, then for $E_n(f,\Psi_a^{(r)},\rho)_{p'}$ they give upper bounds which are exact with respect to order for the corresponding classes $MH(0,0,p,G(b))$.

If $a < b$, for every $|\rho| < \rho(\Phi_a^{(r)}, M)$ the estimates of the remainder terms in (4.14) $(p = \infty)$ *and* (4.15) $(1 \leqq p < \infty)$ *are also exact in the sense of order. They cannot be improved in the corresponding classes $MH_{\rho,\Phi_a^{(r)}}$.*

When $a \neq b$ the statements about accuracy of (4.14) and (4.15) are, as usual, already proved (see Theorem 4.2). If $a = b$, then in the case where $p = \infty$ it is necessary to consider the sequence of functions $Q_n(z) \in MH(0, 0, \infty, G(a))$, defined by formula (4.3), and in the case where $1 \leqq p < \infty$, the sequence of functions $K_n(z) = K_n(z, p) \in MH(0, 0, p, G(a))$, defined by formula (4.10). Assuming that

$$p^*_{n-1}(z) = \sum_{s=1}^{n} \frac{1}{s} \frac{T_{n-s}(z)}{g^{n-s}(a)}$$

and

$$\pi^*_{n-1}(z) = n^{-\frac{1}{q}} \sum_{s=1}^{n-1} \left(1 - \frac{s}{n}\right) \frac{T_{n-s}(z)}{g^{n-s}(a)} \left(\frac{1}{p} + \frac{1}{q} = 1\right),$$

it is easy to calculate that

$$\| Q_n(x) - p^*_{n-1}(x) \|_{L[-1,1]} \leqslant C(M) g^{-n}(a),$$

$$\| K_n(x) - \pi^*_{n-1}(x) \|_{L[-1,1]} \leqslant C(M) g^{-n}(a)$$

and

$$| \Psi_a^{(r)}(p^*_{n-1}) - \rho | \geqslant C n^r \ln n$$

$$| \Psi_a^{(r)}(\pi^*_{n-1}) - \rho | \geqslant C n^{r+\frac{1}{p}} \quad (C > 0,\ n = 1, 2, \ldots).$$

From this and from inequalities (1.8) and (3.6) we obtain our statements about the accuracy of the estimates of (4.14) and (4.15) in the case where $a = b$.

Let us consider the behavior of the best approximations of analytic functions by polynomials with restraints generated by the functional $\Psi_{\alpha,a}^{(r)}$ (see Definition 3.3).

THEOREM 4.6. *Let $f(z) \in MH(k, \beta, p, G(b))$ and let $1 \leqq p' \leqq \infty$. Then, retaining the restrictions and notation of Theorem 3.6, when $n \to \infty$ we have:*

1) *if $p = \infty$, then*

$$E_n(f, \Psi_{\alpha,a}^{(r)}, \rho_n)_{p'} = \frac{|\Phi_{\alpha,a}^{(r)}(f) - \rho_n|}{\|\Psi_{\alpha,a}^{(r)}\|_{n,p'}} + O\left\{\frac{1 + \delta_{a,b} \ln n}{n^{k+\beta} g^n(b)}\right\}; \tag{4.16}$$

2) *if $1 \leqq p < \infty$, then*

$$E_n(f, \Psi_{\alpha,a}^{(r)}, \rho_n)_{p'} = \frac{|\Phi_{\alpha,a}^{(r)}(f) - \rho_n|}{\|\Psi_{\alpha,a}^{(r)}\|_{n,p'}} + O\left\{\frac{1 + \delta_{a,b} n^{1/p}\left(1 + \delta_{r+\frac{1}{p}-\alpha,\, k+1+\beta} \ln n\right)}{n^{k+\beta} g^n(b)}\right\}, \tag{4.17}$$

where the constants entering into O depend only on the quantities M, b, k, β, p and on the functional $\Psi^{(r)}_{\alpha,a}$.

If the parameters a, b, r, k, β, p and $\rho_n \equiv \rho$ *are such that these formulas are not asymptotic, then for* $E_n(f, \Psi^{(r)}_{\alpha,a}, \rho)_{p'}$ *they give upper bounds which are exact with respect to the corresponding classes* $MH(k,\beta,p,G(b))$. *If* $a = b$ *and* $r + 1/p - \alpha < k + 1 + \beta$ *or* $a < b$, *then for every* $|\rho| < \rho(\Phi^{(r)}_{\alpha,a}, M)$ *the estimates of the remainder terms in* (4.16) *and* (4.17) *are also exact in the sense of order. It is impossible to improve them in the corresponding classes* $MH_{\rho,\Phi^{(r)}_{\alpha,a}}$.

Formulas (4.16) and (4.17) are derived from equality (1.6′), Lemma 3.2, Theorem 3.6 and Theorems 2.1, 2.2.

In the case where $a = b$ the statements about accuracy of (4.16) and (4.17) may be obtained by considering the same examples as were presented in the proofs of Theorems 4.3—4.5.

In conclusion, let us present two theorems of the best approximations of $E_n(f, \Psi, \rho_n)_{p'}$ when the sequence of norms $\|\Psi\|_{n,p'}$ of the functional Ψ has exponential growth when $n \to \infty$.

THEOREM 4.7. *Let* $k \geqq 0$ *be an integer; take* $0 < \beta \leqq 1$ *if* $k > 0$ *and* $0 \leqq \beta \leqq 1$ *if* $k = 0$; $M > 0$, $b > 1$, $1 \leqq p \leqq \infty$, $1 \leqq p' \leqq \infty$, $\nu > 0$. *Let the homogeneous additive functional* $\Psi(p_n)$ *satisfy the condition*

$$\|\Psi\|_{n,\,p'} \asymp n^{\nu} \quad (n \to \infty)$$

and let $f(z) \in MH(k,\beta,p,G(b))$. *Then*

$$E_n(f, \Psi, \rho_n)_{p'} = \frac{|\Phi(f) - \rho_n|}{\|\Psi\|_{n,\,p'}} + O\left(\frac{C(M, \Psi)}{n^{k+\beta}\, g^n(b)}\right), \tag{4.18}$$

where $\Phi(f) = \Phi(f;\Psi)$ *is the same functional as in Theorem* 3.7.

Using the same arguments as in the proof of Theorem 4.2, it is possible to demonstrate that in the class $MH_{\rho,\Phi}(k,\beta,p,G(b))$ the estimate given by (4.18) is exact in the sense of order.

Equality (4.18) follows from formula (1.6) and Theorem 3.7.

Let us remember that we have denoted by $Mh(k,\beta,p)$ (see §3) the set of functions $f(x)$ from $L_p[-1,1]$ for which

$$E_n(f)_p \leqslant Mn^{-k-\beta} \; (k \geqslant 0,\ 0 < \beta \leqslant 1,\ 1 \leqslant p \leqslant \infty).$$

THEOREM 4.8. *Let* $k > 0$ *be an integer, let* $1 \leqq p \leqq \infty$, $0 < \beta \leqq 1$, $\nu > 0$, $M > 0$, *and let* $\Psi(p_n)$ *be any homogeneous additive functional satisfying the condition*

$$\|\Psi\|_{n,\,p} \asymp n^{\nu}\,(n \to \infty)$$

Let $f \in Mh(k,\beta,p)$ *and let* $\Phi(f) = \Phi(f;\Psi)$ *be the same functional as in Theorem* 3.8. *Then*

$$E_n(f, \Psi, \rho_n)_p = \frac{|\Phi(f) - \rho_n|}{\|\Psi\|_{n,\,p}} + O\left(\frac{1 + \delta_{\nu,\,k+\beta} \ln n}{n^{k+\beta}}\right) \tag{4.19}$$

where the constant entering into O depends only on the quantities M, k, β, p and on the functional Ψ. If in the right-hand side the principal term is the second term when $n \to \infty$ and $\rho_n \equiv \rho$, then the order of this term in the class $Mh(k,\beta,p)$, generally speaking, cannot be reduced.

Let us explain that the phrase "generally speaking" means the following here: If the order of the right-hand side of formula (4.19) is $n^{-(k+\beta)} \ln n$, then we cannot prove the statement of accuracy of formula (4.19) for any functional Ψ satisfying the conditions of the theorem. However, we can present an example of a functional Ψ for which this is true.

Proof. Formula (4.19) follows from equation (1.6′) and Theorem 3.8. If $\nu \neq k+\beta$, the statement of accuracy of formula (4.19) is trivial.

To prove the accuracy of (4.19) when $\nu = k+\beta$ let us consider the cases where $2 < p \leqq \infty$, $1 \leqq p < 2$ and $p = 2$ separately.

Let $\nu = k+\beta$ and $2 < p \leqq \infty$. Let us define the functional $\Psi_\nu(p_n)$ in the following way. Take $\Psi_\nu(T_n) = 0$ if

$$n \neq 2^j;\ \Psi_\nu(T_{2^j}(x)) = 2^{j\nu}\ (j = 0, 1, \ldots).$$

Then for an arbitrary polynomial

$$p_n(x) = \sum_{l=0}^{n} a_l T_l(x)$$

we have

$$\Psi_\nu(p_n) = \sum_{l=0}^{n} a_l \Psi_\nu(T_l(x)) = \frac{2}{\pi}\int_{-1}^{1} p_n(t) \sum_{l=1}^{n} \Psi_\nu(T_l) T_l(t) \frac{dt}{\sqrt{1-t^2}}.$$

Let us demonstrate that for any p $(2 < p \leqq \infty)$

$$\|\Psi\|_{n,p} \asymp C(p) n^\nu.$$

In fact, if $2^s \leqq n < 2^{s+1}$ and $1/p + 1/q = 1$, then

$$\begin{aligned} 2^{-\nu} n^\nu &< 2^{s\nu} = \Psi_\nu(T_{2^s}) \leqslant \|\Psi_\nu\|_{2^s} \leqslant \|\Psi_\nu\|_n \leqslant \|\Psi_\nu\|_{n,p} \\ &= \sup_{\|p_n\|_{L_p[-1,1]} \leqslant 1} \left| \frac{2}{\pi} \int_{-1}^{1} p_n(t) \left[\sum_{l=1}^{n} \Psi_\nu(T_l) T_l(t)\right] \frac{dt}{\sqrt{1-t^2}} \right| \\ &\leqslant \frac{4}{\pi} \sum_{l=1}^{n} \Psi_\nu(T_l) \|(1-t^2)^{-\frac{1}{2}}\|_{L_q[-1,1]} = C(p) \sum_{j=1}^{s} 2^{j\nu} \\ &\leqslant C(p,\nu) 2^{s\nu} \leqslant C(p,\nu) n^\nu \quad (\nu = k+\beta). \end{aligned}$$

Let us assume that

$$f_\nu(x) = \sum_{l=1}^{\infty} 2^{-l\nu} T_{2^l}(x).$$

For all p $(2 < p \leqq \infty)$ and n $(2^{s-1} \leqq n < 2^s)$

$$E_n(f_\nu)_p \leqslant E_n(f_\nu) \leqslant \left\| \sum_{l=s}^{\infty} 2^{-l\nu} T_{2^l}(x) \right\|_{C[-1,1]} = \frac{M}{2^{s\nu}} \leqslant \frac{M}{n^{k+\beta}},$$

that is, $f_\nu(x) \in Mh(k,\beta,p)$. On the other hand, on the basis of inequality (1.8) we have

$$E_n(f_\nu, \Psi_\nu, \rho)_p \geqslant \|\Psi_\nu\|_{n,p}^{-1} \left| \Psi_\nu\left(\sum_{l=0}^{s-1} 2^{-l\nu} T_{2^l}(x)\right) - \rho \right|$$

$$- \left\| f_\nu(x) - \sum_{l=0}^{s-1} 2^{-l\nu} T_{2^l}(x) \right\|_{L_p[-1,1]} > Cn^{-\nu} \sum_{l=0}^{s-1} 1 + O\left(\frac{\rho}{n^\nu}\right)$$

$$= Cn^{-\nu}s + O\left(\frac{\rho}{n^\nu}\right) = Cn^{-\nu} \ln 2^s + O\left(\frac{\rho}{n^\nu}\right) > C \ln n\, n^{-k-\beta} \quad (C > 0),$$

if ρ does not depend on n. Q.E.D.

If $1 \leq p < 2$, then assuming $\Psi_\nu(2^{-l}T'_{2^l}(x)) = 2^{l\nu}$, $\Psi_\nu(T'_n(x)) = 0$ if $n \neq 2^l$ $(l = 0, 1, \cdots)$, and

$$f_\nu(x) = \sum_{l=0}^{\infty} 2^{-l(\nu+1)} T'_{2^l}(x) \quad (\nu = k + \beta),$$

using analogous arguments we can demonstrate that

$$\|\Psi_\nu\|_p \asymp C(p)\, n^\nu \quad (1 \leqslant p < 2)$$

and

$$E_n(f, \Psi_\nu, \rho)_p > Cn^{-k-\beta} \ln n \quad (C > 0).$$

If $p = 2$, then we can take as an example

$$f(x) = \sum_{l=0}^{\infty} 2^{-l\nu} L_{2^l}(x),$$

where $\{L_n(x)\}$ is a system of Legendre polynomials which is orthonormalized in the interval $[-1,1]$, and the functional $\Psi_\nu(p_n)$ satisfies the conditions $\Psi_\nu(L_{2^l}) = 2^{l\nu}$ $(\nu = k + \beta)$, and

$$\Psi_\nu(L_n) = 0 \quad \text{if} \quad n \neq 2^l \quad (l = 0, 1, \ldots).$$

In conclusion let us note that the results obtained here may be carried over to the case of approximation of functions by trigonometric polynomials. Articles [9, 10] and [14] contain some results in this area.

BIBLIOGRAPHY

1. V. A. MARKOV, *Functions deviating least from zero*, St. Petersburg, 1892. (Russian)
2. V. I. SMIRNOV AND N. A. LEBEDEV, *The constructive theory of functions of a complex variable*, "Nauka", Moscow, 1964. (Russian) MR **30** #2152.
3. S. M. LOZINSKIĬ AND I. P. NATANSON, *Metric and constructive theory of functions of a real variable*, in *Forty Years of Mathematics in the USSR*: 1917-1957. Vol. 1: *Survey articles*, Fizmatgiz, Moscow, 1959, pp. 363-367. (Russian) MR **22** #6672.

4. B. A. RYMARENKO AND V. N. BUROV, *Some conditional extremal problems*, Proc. Fourth All-Union Math. Congress. Vol. II, "Nauka", Moscow, 1964, pp. 680-683. (Russian) MR **29** #4649.

5. S. JA. HAVINSON, *Extremal and approximation problems with additional conditions*, in *Studies of modern problems of the constructive theory of functions*, Fizmatgiz, Moscow, 1961, pp. 345-352. (Russian) MR **33** #480.

6. V. N. BUROV, *On two types of conditional extremum problems and a general approach to their investigation*, Ukrain. Mat. Ž. **17** (1965), 107-111. (Russian) MR **32** #8009.

7. N. I. ČERNYH, *Approximation of functions by polynomials whose coefficients are connected by a linear dependence*, Proc. Second Sci. Conf. Math. Dept. Ped. Inst. Volga Region, no. 1, Kuĭbyšev. Gos. Ped. Inst., Kuybyshev, 1962, pp. 113-117. (Russian) MR**34** #1760.

8. ———, *Some extremal problems for polynomials*, Trudy Mat. Inst. Steklov. **78** (1965), 48-89 = Proc. Steklov Inst. Math. **78** (1965), 49-90. MR **32** #7690.

9. M. ZAMANSKY, *Sur l'approximation des fonctions continues*, C. R. Acad. Sci. Paris **226** (1948), 1066-1068. MR **9**, 582.

10. S. B. Stečkin, *On the order of the best approximations of continuous functions*, Izv. Akad. Nauk SSSR Ser. Mat. **15** (1951), 219-242. (Russian) MR **13**, 29.

11. D. JACKSON, *Über Genauigkeit der Annäherung stetiger Functionen durch ganze rationale Functionen gegebener Grades und trigonometrische Summen gegebener Ordnung*, Dissertation, Gottingen, 1911.

12. N. I. AHIEZER, *Lectures on the theory of approximation*, OGIZ, Moscow, 1947; 2nd rev. ed., "Nauka", Moscow, 1965; English transl., Ungar, New York, 1956. MR **3**, 234; MR **20** #1872; MR **32** #6108.

13. J. L. WALSH AND H. G. RUSSELL, *Integrated continuity conditions and degree of approximation by polynomials or by bounded analytic functions*, Trans. Amer. Math. Soc. **92** (1959), 355-370. MR **21** #7311.

14. N. I. ČERNYH, *Approximation of functions by polynomials with linked coefficients*, Dokl. Akad. Nauk SSSR **162** (1965), 290-293 = Soviet Math. Dokl. **6** (1965), 679-683. MR **31** #5018.

15. L. A. LJUSTERNIK AND V. I. SOBOLEV, *Elements of functional analysis*, GITTL, Moscow, 1951; 2nd rev. ed., "Nauka", Moscow, 1965; English transl., Hindustan, Delhi, Gordon & Breach, New York, Ungar, New York, 1961. MR **14**, 54; MR **25** #5362; MR **35** #698.

16. S. N. BERNŠTEĬN, *Extremal properties of polynomials and best approximation of continuous functions of a real variable*, Glaz. Redek. Obšč. Lit., Leningrad, 1937. (Russian)

17. I. I. PRIVALOV, *Boundary properties of analytical functions*, GITTL, Moscow, 1950; German transl., Hochschulbücher für Math., Bd. 25, VEB Deutscher Verlag, Berlin, 1956. MR **13**, 926; MR **18**, 727.

18. A. I. MARKUŠEVIČ, *The theory of analytic functions*, GITTL, Moscow, 1950; English transl., Prentice-Hall, Englewood Cliffs, N. J., 1965. MR **12**, 87; MR **31** #5965.

19. A. A. MARKOV, *On a problem of D. I. Mendeleev*, Izv. Akad. Nauk, St. Petersburg **62** (1899), 1-24; Reprinted in *Selected works on continued fractions and the theory of functions deviating least from zero*, GITTL, Moscow, 1948, pp. 51-75. (Russian)

20. P. L. ČEBYŠEV, *Functions differing little from zero for certain values of the variable*, Zap. Imp. Akad. Nauk **40** (1881), no. 3; reprinted in *Complete collected works*. Vol. III, Izdat. Akad. Nauk SSSR, Moscow, pp. 108-127; French transl., in *Oeuvres de P. L. Tchebychef*. Vol. II, St. Petersburg, 1907. MR **11**, 150.

21. N. K. BARI, *Generalization of inequalities of S. N. Bernšteĭn and A. A. Markov*, Izv. Akad. Nauk SSSR Ser. Mat. **18** (1954), 159-176. (Russian) MR **15**, 788.

22. S. B. STEČKIN, *Generalization of some inequalities of S. N. Bernšteĭn*, Dokl. Akad. Nauk SSSR **60** (1948), 1511-1514. (Russian) MR **9**, 579.

23. N. K. BARI, *Trigonometric series*, Fizmatgiz, Moscow, 1961. (Russian) MR **23** #A3411.